QUARKS, CRITTERS, AND CHAOS

5/93

QUARKS, CRITTERS, AND CHAOS

WHAT SCIENCE TERMS REALLY MEAN

JO ANN SHROYER

PRENTICE HALL GENERAL REFERENCE
New York London Toronto Sydney Tokyo Singapore

To my family and friends—especially my husband, J. G. Preston, and my parents, Beryl and Frances Shroyer.

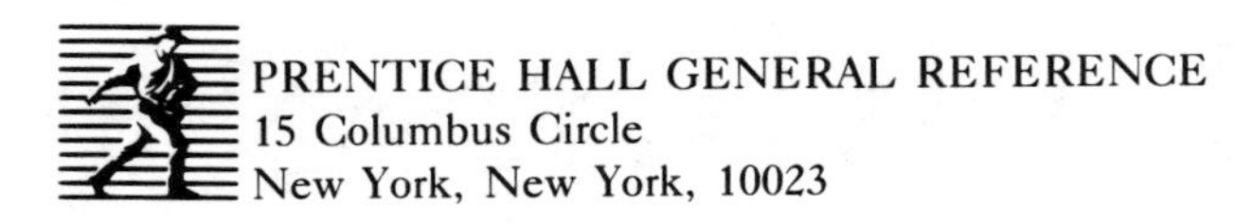
PRENTICE HALL GENERAL REFERENCE
15 Columbus Circle
New York, New York, 10023

Library of Congress Cataloging-in-Publication Data

Shroyer, Jo Ann.
Quarks, critters, and chaos : what science terms really mean / Jo Ann Shroyer.
p. cm.
Includes bibliographical references and index.
ISBN 0-671-84744-9—ISBN 0-671-84745-7 (pbk.)
1. Science—Terminology. 2. Science—Popular works.—I. Title.
Q123.S519 1993
501.4—dc20 92-22614
CIP

Design and illustrations by Irving Perkins, Associates, Inc.

Manufactured in the United States of America

10 9 8 7 6 5 4 3 2 1

First Edition

ACKNOWLEDGMENTS

I am grateful to all those scientists who talked with me and answered my questions. I am especially indebted to those researchers and other specialists who read portions of my manuscript and made helpful suggestions: Larry Rubin, Ken Taylor, William Thompson, Scott McIvor, Jeffrey Tate, Maria Gini, Robert Sloan, Vipin Kumar, Betsy Perry, Earl Peterson, Shashi Shekhar, Ralph Huessner, Miron L. Heinselman, Richard Skaggs, Tom Vandevander, Evan Skillman, Lynn Solem, Mark Pittlekow, Germaine Cornelissen, Darin O'Brien, and Judy Schanke.

Thanks to the Science Museum of Minnesota's *Encounters* magazine, where some of these articles were first published.

I am enormously grateful to my agent, Jeanne Hanson, for her steadfast guidance and friendship, to my dear friends, Patricia Neuman and Linda Phipps, for reading parts of my manuscript and helping me in many other ways, and to my children, Peter, Claire, Jessica, and Katherine, for tolerating my preoccupation with work. Most of all, I am grateful to J. G. Preston, my husband, for his unselfish and enthusiastic support.

CONTENTS

PREFACE

Ours is an age of tremendous change in technology and science. Every week something new appears in our newspapers and on radio and television. Reporters toss around words like *black hole*, *genetic engineering*, *artificial intelligence*, *antimatter*, and *chaos*, and try to explain what these terms mean to scientists and also to the rest of us, whose lives may very well be affected by the results of such research. Writers try to interpret the findings of scientists who spend their lives carefully unfolding the mysteries that inhabit every aspect of our world and our universe. But sometimes they don't make sense.

This should not be a surprise. Cosmology, subatomic physics, genetics—these are difficult subjects. Nevertheless, discoveries in these and other disciplines may have long-range effects on our lives. We sometimes have to make decisions based upon research in these fields. How far should geneticists take their capability of manipulating the genes of living things, including humans? How much time and money should we spend on finding the fundamental particles of matter? Why do we need expensive, high-tech telescopes to probe the secrets of the universe?

We often long for practical applications for such research and ask, "What good is this going to do in a world with so many down-to-earth needs?" There may not be practical applications for every discovery, of course. But we are sometimes surprised. Research in physics resulted in solid state electronics—a practical application that has affected almost everyone. The electronics revolution that gave us computers, compact discs, and cellular telephones could not have occurred without physics research. Discoveries in high energy physics gave doctors diagnostic tools that can view the workings of the living brain and find evidence of cancer and other diseases. The scientists who figured out how atoms were arranged in the DNA molecule began a mapping process that makes possible the intervention in genetic disease and the laboratory production of vital human hormones like insulin.

We are also curious to know the answers to questions that have niggled at humans for generations. How did our universe begin? What lies beyond the beyond? How did the Earth and all the diverse life upon it come to be? What killed the dinosaurs and how can we save ourselves and other species from extinction?

The answer to these and many other questions can be found in science, along with some really good stories, too. One researcher remarked to me, "Who needs fiction? There are enough mysteries, good stories, and amazing situations in the study of the natural world to satisfy anyone!" Science, after all, is more than formulas, theories, and dry, carefully worded reports. The universe and our small part of it is filled with strange objects and beings, secret patterns, violence, beauty, and adventure. The study of these wonders is painstakingly careful, while at the same time it is infused with intuition, poetry, dreaming, and competition.

The problem, of course, is communicating those stories in a language that ordinary people can understand. Not all scientists are interested in reducing their work to simple language and comfortable analogies. Others feel strongly that everyone should have at least some understanding of the processes and discoveries that help us to know ourselves and our universe better and to make intelligent decisions when necessary.

This book is for all those who wish to have such a speaking acquaintance with the scientific terms they see and hear in the media. It is also for those who have at one time or another felt that science is a secret language, a private party that excludes the ordinary person because of its complexity and the specialized nature of its study.

Science is not an exclusive club. It's the story of our world, our universe, close in and far out. It's the beating of our hearts, the stems of grass beneath our feet, the earth that sifts through our fingers. Science is all around us, and we are privileged and obligated to understand it and share in its wonders.

I

ATOMS, MOLECULES, AND STARDUST MEMORIES

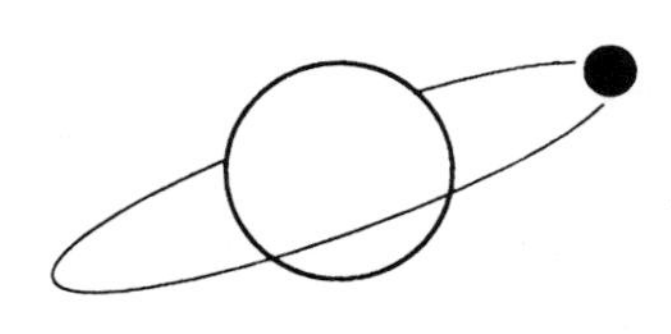

SCIENTISTS BELIEVE that the universe began with a bang and evolved from a seething soup pot of particles that coalesced into atoms that congregated into molecules. As time went on, matter clumped and clustered into a honeycomb of gas clouds, bright hot stars, galaxies, and groups of galaxies. Insignificant clots of dust and all the elements from which life could spark were gathered into planets. The elements from which we and everything on Earth are made were forged in the fiery furnace of the stars from raw material created at the beginning of the universe.

Physicists study the components of this universe on the very smallest scale, exploring the fundamental particles of matter and energy and their interactions. Astrophysicists, astronomers, and cosmologists look far into space and back in time to find out how those particles clustered together to create a universe in which whirling disks of dusty gas became bright stars, black holes, and tiny planets.

Scientists believe that the forces that control all of these wonders were given expression during the first moments after the Big Bang and then split apart and set about their own business. But there is no proof, and one of the great goals of modern physics is to develop theories that will unify the forces that describe the very smallest and the very largest structures we know. It turns out they can't investigate one without understanding the other.

Nevertheless, it's amazing to think that by looking deep into the secret world of atoms and molecules, scientists have been able to understand the vast workings of the universe, and that by exploring space, they have found that most of the rules that govern the heavens apply on the smallest scale as well.

ATOMS

Atoms are the basic building blocks of matter. Everything, including the ink in these printed words, is made from atoms. For centuries, scholars suspected their existence but had no way of proving it. The notion was probably first suggested more than two thousand years ago by Democritus, a Greek scholar who gave it a name—*atomos*, which means "unbreakable."

The atom is the smallest particle of an element that can take part in a chemical reaction without being permanently changed. Seventeenth-century British chemist Robert Boyle (1627–91) began assembling proof for the existence of atoms by identifying all the substances which could not be broken into simpler parts. Since then, scientists have worked to identify all the elements that make up our world and the parts of the universe that can be studied.

There are now more than one hundred chemical elements, which are arranged in the periodic table shown at right. Each box in the table provides an element's symbol, mass, and atomic number, which increases as you read to the right. The atomic number reflects the number of protons in an atom's nucleus. For example, hydrogen, the simplest of elements, has one proton in its nucleus, and thus an atomic number of 1. The atomic mass, indicated by a number at the bottom of each block in the periodic table, reflects the number of neutrons in the nucleus. In the periodic table, mass refers to the amount of matter of a particular element in an atom relative to an agreed upon standard of measurement. The current standard is the atomic mass of Carbon–12, the most abundant isotope of carbon. An isotope is a form of an element that has the same atomic number and chemical properties as all forms of that element, but has a different atomic mass because of a differing number of neutrons in the nucleus.

It was John Dalton, a nineteenth-century English chemist (1766–1844), who gave us our modern theory of the atom, resurrecting the term coined by Democritus. His experiments showed that atoms, representing different elements, combined in many different ways to make compounds, but no matter how they combined with others, the individual atoms maintained a fixed mass and identity.

Scientists continued to do experiments aimed at unlocking the secrets of the atom, arguing over whether atoms really existed or were just a convenient, descriptive tool. Then in 1905, Albert Einstein worked out mathematical equations that supported the existence of the atom.

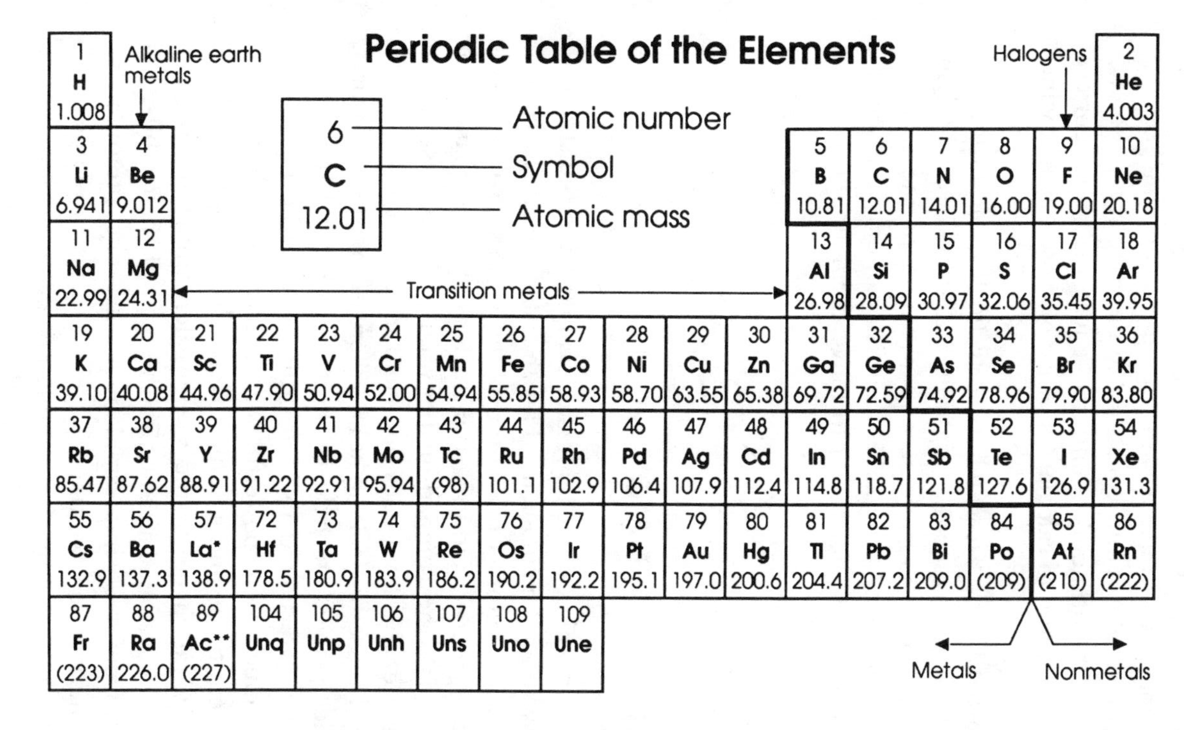

Periodic Table of the Elements

1 **H** 1.008																	2 **He** 4.003
3 **Li** 6.941	4 **Be** 9.012											5 **B** 10.81	6 **C** 12.01	7 **N** 14.01	8 **O** 16.00	9 **F** 19.00	10 **Ne** 20.18
11 **Na** 22.99	12 **Mg** 24.31											13 **Al** 26.98	14 **Si** 28.09	15 **P** 30.97	16 **S** 32.06	17 **Cl** 35.45	18 **Ar** 39.95
19 **K** 39.10	20 **Ca** 40.08	21 **Sc** 44.96	22 **Ti** 47.90	23 **V** 50.94	24 **Cr** 52.00	25 **Mn** 54.94	26 **Fe** 55.85	27 **Co** 58.93	28 **Ni** 58.70	29 **Cu** 63.55	30 **Zn** 65.38	31 **Ga** 69.72	32 **Ge** 72.59	33 **As** 74.92	34 **Se** 78.96	35 **Br** 79.90	36 **Kr** 83.80
37 **Rb** 85.47	38 **Sr** 87.62	39 **Y** 88.91	40 **Zr** 91.22	41 **Nb** 92.91	42 **Mo** 95.94	43 **Tc** (98)	44 **Ru** 101.1	45 **Rh** 102.9	46 **Pd** 106.4	47 **Ag** 107.9	48 **Cd** 112.4	49 **In** 114.8	50 **Sn** 118.7	51 **Sb** 121.8	52 **Te** 127.6	53 **I** 126.9	54 **Xe** 131.3
55 **Cs** 132.9	56 **Ba** 137.3	57 **La*** 138.9	72 **Hf** 178.5	73 **Ta** 180.9	74 **W** 183.9	75 **Re** 186.2	76 **Os** 190.2	77 **Ir** 192.2	78 **Pt** 195.1	79 **Au** 197.0	80 **Hg** 200.6	81 **Tl** 204.4	82 **Pb** 207.2	83 **Bi** 209.0	84 **Po** (209)	85 **At** (210)	86 **Rn** (222)
87 **Fr** (223)	88 **Ra** 226.0	89 **Ac**** (227)	104 **Unq**	105 **Unp**	106 **Unh**	107 **Uns**	108 **Uno**	109 **Une**									

*Lanthanides (Rare Earths)	58 **Ce** 140.1	59 **Pr** 140.9	60 **Nd** 144.2	61 **Pm** (145)	62 **Sm** 150.4	63 **Eu** 152.0	64 **Gd** 157.3	65 **Tb** 158.9	66 **Dy** 162.5	67 **Ho** 164.9	68 **Er** 167.3	69 **Tm** 168.9	70 **Yb** 173.0	71 **Lu** 175.0
**Actinides (Transuranium)	90 **Th** 232.0	91 **Pa** (231)	92 **U** 238.0	93 **Np** (237)	94 **Pu** (244)	95 **Am** (243)	96 **Cm** (247)	97 **Bk** (247)	98 **Cf** (251)	99 **Es** (252)	100 **Fm** (257)	101 **Md** (258)	102 **No** (259)	103 **Lr** (260)

An 80-year-old hypothesis suggested that the vibrations of pollen grains in water were caused by the bumping and jostling of atoms in the water molecule. Einstein assumed that something real was making this happen and wrote equations to describe the phenomenon. Subsequent experiments found the equations to be correct, and scientists used them to calculate the size of these objects, which no one had ever seen.

It wasn't until 1955 that researchers were finally able to magnify a substance sufficiently to see the bright little dots that were individual atoms. Now, with the more recent invention of the scanning tunneling microscope (see page 231), scientists not only can see and photograph single atoms, but they can move them around like golf balls.

THE STRUCTURE OF THE ATOM

The easiest way to think of an atom is to compare it to the structure of the solar system. The nucleus, like the sun, is at the center, contains most of the atom's mass, and is the repository of a prodigious amount of energy. Orbiting the nucleus are tiny planets, or electrons. Although the comparison between the solar system and the construction of the atom does not stand up to much scrutiny, it is a jumping-off place for looking at an otherwise mysterious world.

For example, planets are held in their orbits around the sun by gravity. Electrons are bound to their orbits around the nucleus by a force that is much more powerful than gravity—the electrical attraction between them and the positively charged nucleus. The amount of electrical charge in the nucleus determines the number of electrons in orbit around it and also the radii of their orbits. For example, hydrogen, the simplest element, has one proton in its nucleus and one electron orbiting it. Helium has two protons, two electrons, and a tighter orbit. The strong nuclear attraction of additional protons in the nucleus draws in the inner orbits so that the heavier elements in the periodic table are not much bigger than the lighter elements.

Planets are separated from each other and the sun by billions of miles. Within the atom, the nucleus and electrons are separated by distances that are vast relative to the size of the particles. As a result, the material that is built of these atoms consists almost entirely of empty space, even though it seems quite solid to us. Fortunately, the like electrical charges of the atoms in matter repel each other. Otherwise, objects would pass through each other like ghosts walking through a wall, posing mind-boggling difficulties.

So, a simple comparison between the atom and the solar system soon breaks down, because the hidden world of the atom follows a different set of rules called quantum theory.

First suggested by physicist Max Planck in 1900, quantum theory proposes that energy is emitted in discrete quanta or units, rather than what it seems to us—a continuous flow. The name for this theory derives from a Latin root that means "how much." Scientists have come to refer to the body of rules as the standard model. The quantum is a measurement of action, and on the atomic level it's an all-or-nothing proposition.

ELECTRONS

Physicists have separated out a remarkable zoo of particles and strange forces in the atom, and there is little in our everyday lives to which we can compare it.

For example, the electron or planet in our tiny, imaginary solar system can move from one orbit to another without having spent any time in between the orbits. Something that was here in one orbit is suddenly there in another, leaving no evidence of the journey. This is called a quantum leap. If planets in the solar system behaved in the same way, things would be in a terrible mess.

It's the electron's leap from orbit to orbit, however, that accounts for the properties of matter and most of the energy transactions in life. When an electron leaps to an orbit that is farther away from the nucleus of the atom, it absorbs energy. When it drops to a lower

Energy Levels of an Atom

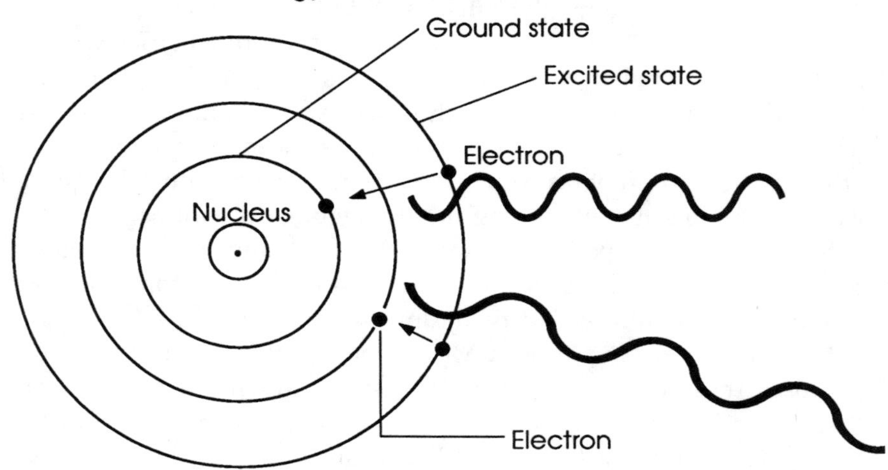

orbit, it sheds energy in the form of little radiant packets of light called photons.

It is this energy transfer in the atom that makes life possible. Chemical reactions are just rearrangements of electrons within and among atoms. The mortar that holds these little bricks together is the attraction between positive and negative electrical charges. All the variety we see around us is due to the many ways different atoms bond chemically to each other.

When comparing electrons to planets it is tempting to think of them as little spheres or particles, when actually they are both particles and waves, but not in the same sense as a grain of sand or a wave on the sea. We must think of them as something completely different—entities that display the properties of either particle or wave, depending on how you're looking at them. Scientists have observed that electrons can be reflected like particles and can also be interfered with in the way that one wave modifies another.

Other subatomic particles, indeed all things, express the same wave particle duality. But the greater a thing's mass, the more it will act like a particle. Golf balls may be subject to the wave-particle duality; we just aren't able to detect it, nor can we blame it for a bad game.

Electrons spin on an axis somewhat like planets, and just as planets are subject to the push and pull of gravity, electrons respond to magnetic fields. The energy of an electron changes with the direction and speed of its spin—which can be switched back and forth by magnetic force.

Given all this weird behavior, it is comforting to know that electrons, like golf balls, cars, or people, cannot, no matter how hard we try, be forced to occupy the same spot at the same time. An electron's position in an orbit or shell is determined by its level of energy—the number and frequency of its wavelengths as it orbits the nucleus—in addition to its average distance from the nucleus, and the direction of its spin. These characteristics are referred to as quantum number. No two electrons can have the same quantum number. This is called the exclusion principle, and it governs the number of electrons that can occupy any orbit, each of which represents a level of energy. The number of electrons and the way they are arranged in the orbits (or shells) determine the manner in which one atom combines with others to make molecules.

PROTONS AND NEUTRONS

The nucleus is the concentrated nut at the center of an atom. Even though it is extremely small, it weighs over 2,000 times more than an electron. It carries a positive charge that must equal the negative charge of the electrons whizzing around it in order for the atom to be neutral.

The positive charge is contributed by the proton, one of the two kinds of particles that constitute the densely packed nucleus. The proton's partner in the nucleus is the neutron; it has the same mass as the proton but carries a neutral charge. Its function is to hold protons in the nucleus together. Hydrogen, which has only one proton and no neutrons at all, is the simplest element, while the heaviest atoms have as many as 140 neutrons.

Unlike electrons, which seem to be elementary particles and can't be broken down any further, protons and neutrons are made of even smaller parts. Tremendous force is required to blow apart a proton, which is a lucky thing. If protons weren't so stable,

Parts of an Atom

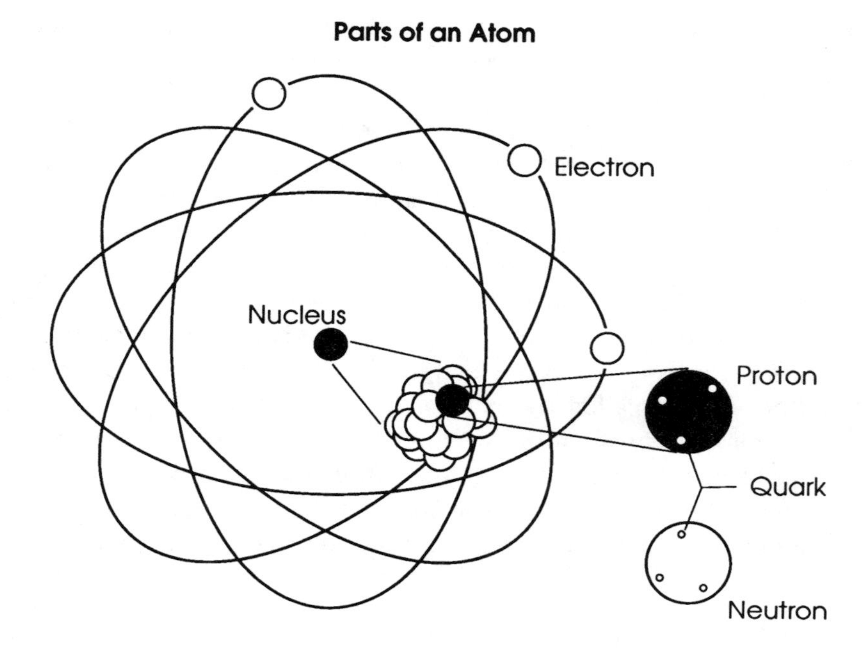

atoms—and everything that's made of them—would be in big trouble.

Despite speculation that protons may have lifetimes as long as the age of the universe, researchers are trying to catch the rare proton in the act of decaying. They have surrounded massive amounts of material—either iron or special fluids—with detectors that will note the flash of proton decay. These experiments are carried out in deep underground caves, tunnels, and mines away from the intrusion of cosmic rays, which could product the same effect.

Our discoveries about atoms, as well as about the workings of the universe, would not have been possible without certain important tools.

ELECTROMAGNETIC SPECTRUM

When you look up at the stars, you are, in effect, looking back in time as well as out into space. Light travels 186,000 miles per second, a distance roughly equivalent to seven trips around Earth at the equator. The starlight we see tonight has traveled for many years to get here. The distances are so vast that scientists measure them in light years, or how far light can travel in a year. Alpha Centauri, our nearest neighboring star, is almost four and a half light years away. When we see Alpha Centauri, we are seeing it as it was four and a half years ago. The Milky Way, an average-size galaxy, is more than 100,000 light years across and is only one of billions and billions of galaxies spread out over impossible distances in the universe.

As astronomers' tools become more powerful, they are able to look farther and farther out in the universe and, of course, back in time to measure light and other radiations from the heavens.

Light from our own star, the sun, takes just a few minutes to travel to us over some 93 million miles. And, just like the light from a reading lamp or any other source, that sunshine is composed of massless, dimensionless packets of radiant energy, or photons, chugging along in trains of waves. Electrons, whirling around the nucleus of the atom, lose energy by spitting out photons and slipping down to lower and lower orbits. Since these packets of energy have no mass, they can zip along forever at the universal speed limit of 186,000 miles per second . . . until they hit something, at which time they disappear. Strange. Very strange.

But the most useful thing to remember about photons is that they carry waves of energy that are measurable. These waves, called radiation, are classified according to their length or frequency, the number of waves per second. The shorter the wave, the more there are per second, and thus, the higher the frequency. On the electromagnetic spectrum, of which visible light is only a small part (less than a millionth of 1 percent), the frequency of waves ranges from gamma rays, with the shortest wave and highest frequency, through X rays, ultraviolet rays, visible light, and infrared rays to the longest, radio wave.

Electromagnetism—one of the fundamental forces that shapes the universe—holds atoms and molecules together by means of the mutual attraction between negative and positive charges. It is also responsible for the waves of energy emitted by these basic building blocks of matter when electrons leap from one orbit to another. The frequency of the waves is related to the energy level of the excited electron.

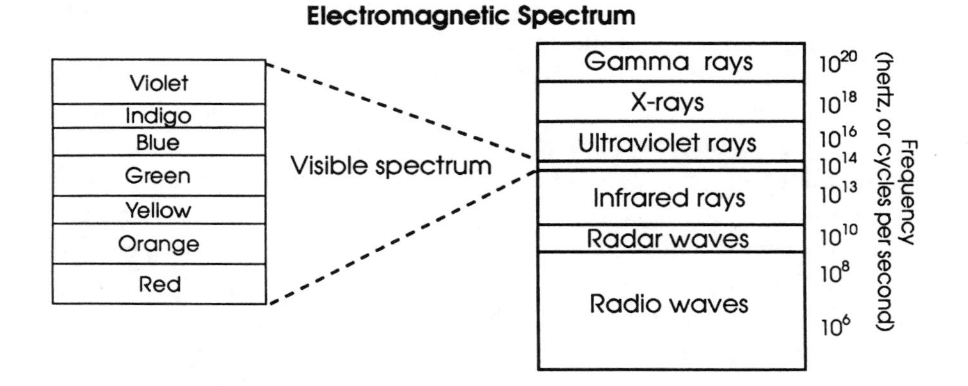

Using tools that detect these various wavelengths of energy, astronomers have been able to study the beams of light and other radiation that shoot from stars, galaxies, clouds of gas, and the many other odd things that seem to be out there. With these tools it is possible to measure and map the universe.

SPECTROSCOPY

In 1665, the British natural philosopher Isaac Newton (1642–1727) discovered that when light shines through a triangular column of glass called a prism, it splinters into a rainbow or spectrum of colors ranging from violet to blue, green, yellow, orange, and red. The colors represent various wavelengths, from high-frequency violet and blue to the lower frequency reds.

The atoms of different elements, when heated, emit different colors or wavelengths of light. It is this characteristic that gives color to every Fourth of July fireworks display. Each element is as individualistic as a fingerprint; it burns with a distinctive spectrum—lines of bright bands superimposed by dark lines, rather like a colorful bar code. Newton's discovery of the light spectrum in 1665 was refined by other scientists, particularly English physicist William Wollaston (1766–1828). In 1802 Wollaston revealed dark lines in the spectrum by passing the light through a narrow slit before it was refracted by the prism. Later a young Bavarian optician, Joseph von Fraunhofer (1787–1826) mapped scientific sequences of these parallel lines in the light emitted by the sun and that reflected by the moon and planets. Two German chemists—Robert Wilhelm Bunsen (1811–99) (who invented the bunsen burner) and Gustav Robert Kirchhoff (1824–87) discovered what these Fraunhofer lines, as they came to be called, really meant; they represented the elements present on Earth in the sun. Their discoveries inspired British astronomer William Huggins (1824–1910) to attach a spectroscope to his telescope. His discovery that the same elements exist in other stars besides the sun opened a new doorway to understanding the universe.

Glowing gas from a burning star has a spectrum, a sort of cosmic recipe, that identifies its temperature and ingredients. Now scientists believe that stars are actually factories for all the elements in the universe. Using spectroscopy, astrophysicists are able to figure out the density, pressure, and temperature of a star, as well as the speed at which it is burning, the strength of its gravity and magnetic field, and the patterns of turbulence on its surface. Modern spectroscopy has grown beyond the dissection of visible light to the study of the whole electromagnetic spectrum.

Spectroscopes have been combined with other technology to interpret and record spectra. Spectrographs make a photographic or digital record of spectra, a spectrometer measures values such as wavelengths of the radiation and energy levels of the particles, and a spectrophotometer is used to compare two spectra with each other. Early spectroscopes were simple devices that grew in size and complexity over the years and now, like other electronic devices, have become more compact. In chemistry labs, they are boxes that fit on a table top. Spectroscopy has given us a better understanding of the life of a star, the workings of the cosmos, and our place in it all.

RED SHIFT AND DOPPLER EFFECT

The spectrum or colorful fingerprint of a star also reveals the speed and direction of its travel. Scientists made these discoveries when they observed that the light from distant galaxies seemed to fall toward the red end of the spectrum. That is, the wavelengths appeared to be slightly longer and, therefore, less frequent. The stars' light contained the same shopping list of elements that local stars contained, but they were slightly shifted in frequency toward the red.

In the 1920s, the American astronomer Edwin Hubble (1889–1953) figured out what was going on. As he was trying to prove that there was more to the universe than our little galaxy, he realized that not only were there other galaxies out there but also that they were moving quickly away from ours and each other. The universe is expanding, he concluded, and the farther away other galaxies are, the faster they are traveling, at a rate directly proportional to their distance. This is called Hubble's Law, and the ratio between speed of motion and distance is referred to as Hubble's Constant.

Hubble explained the red shift problem by calling on the work of nineteenth-century Austrian physicist Christian Johann Doppler (1803–53). Doppler had discovered that the frequency of light waves (and other wave phenomena including sound) increases as it approaches you and decreases as it moves away. You can observe such frequency changes in sound waves; when a wailing train approaches, you can hear the pitch of the whistle rise as the train comes nearer and then fall as it speeds away. This is called the Doppler Effect.

With this tool, scientists were able to see that the stars in our galaxy are not fixed in space but are moving in a jumbly sort of rotation, once around every 230 years. And the knowledge that the universe is expanding at a measurable rate has allowed researchers to measure and map the visible contents of the universe.

Changes in the frequency of light are not detectable with the naked eye, but a telescope fitted with a spectrometer can see them. Until recently, astronomers recorded the red shift of stars and galaxies on photographic plates. Now they use modern electronic detectors that are much more powerful and efficient, recording and detecting 50 to 80 percent of the light that hits them.

INTERFEROMETRY

For hundreds of years astronomers studied the visible stars using only optical telescopes. But many of the objects in the far reaches of outer space are not visible at all. Many of them, however, including the hydrogen atoms that fill interstellar space, emit the longer radio wavelengths that can be detected by special antennae or dishes. The past fifty years have seen the development of telescopes that can detect the nonvisible wavelengths on the electromagnetic spectrum. Radio and X-ray telescopes, gamma-ray detectors, and infrared telescopes have all made it possible to detect things the existence of which scientists had suspected but could not prove.

For example, radio astronomers identified quasars, the brightest objects in the universe, as something other than the stars for which they were mistaken. Unlike stars, quasars are strong sources of radio signals and emit a tremendous burst of light from a relatively small space. (We'll see more about them on page 34.)

In order to get clear images of distant radio sources, the diameter of the telescope dish must be enormous, but beyond a certain point, the size becomes impractical, collapsing under its own weight. Such huge dishes could not be moved and aimed at different parts of the sky, but would be limited to only a small wedge. So at first, radio telescopes were not able to provide clear distinctions between closely spaced objects; the naked eye would be capable of doing better at night than a 150-mile-wide radio telescope dish, if such a thing could exist.

Then in the early 1950s, radio astronomers found a way to create a huge dish of sorts, by bringing together the signals of two radio telescopes, creating the equivalent of a radio dish with a diameter equal to the distance between the two telescopes. These telescopes were linked at first by cable and later by radio. Today, Very Long Base Interferometry utilizes widely separated telescopes that record observations simultaneously on separate magnetic tapes, coordinated by time signals from an atomic clock. Later, one recording is superimposed on the other to reveal a much clearer image. It is now even possible to make a radio dish with a diameter greater than Earth's by using a telescope on a satellite as it orbits our planet.

HUBBLE SPACE TELESCOPE

Despite the improvements made in telescopes, there was still one big problem. Like a dirty windshield, Earth's atmosphere kept astronomers from seeing where they were going. But the Hubble Space Telescope (HST), launched into orbit in April 1990, despite disappointing flaws, has allowed scientists a clearer view of the universe.

Named after astronomer Edwin Hubble, HST was almost twenty years in the making. It was first suggested by a German rocket scientist, who recognized the benefits of placing a telescope beyond the distortions caused by turbulence in the atmosphere. Advances in ground-based telescopes could never solve that problem, or overcome the fact that the atmosphere blocked out portions of the electromagnetic spectrum. Ultraviolet light, for example, is blocked by ozone in the stratosphere, and astronomers have gotten around this obstacle by using detectors on satellites. HST, the most powerful observational device to be sent into orbit, is designed to observe ultraviolet radiation as well as visible light emissions.

HST is 43 feet long and weighs 25,500 pounds. Launched by space-shuttle astronauts, HST is designed to be serviced on the shuttle and even taken back to Earth for repairs, if necessary. Wrapped in Mylar plastic like an oddly shaped party balloon, HST orbits the Earth every hour and a half. It carries two cameras that can see extremely faint objects, two spectrographs for analyzing the chemical makeup of stars and other objects, and a device that measures brightness. This long-awaited telescope is designed to turn quickly and focus accurately on a target, remaining focused on it for up to 24 hours. Costing $1.5 billion, HST has been plagued by problems, to the disappointment of scientists and taxpayers. The telescope developed a case of the jitters, which scientists attributed to the expansion and contraction of loose solar panels. And one of the two spectrographs failed because of what is thought to be a faulty electrical connection. Most distressing of all was the revelation that the 94-inch silica mirror was apparently not ground to the right curvature. HST has not been able to deliver the vastly improved resolution that was a major reason for its development.

On the plus side, HST has allowed scientists some interesting views of the universe, including the clearest pictures that have been taken to date of Mars from Earth's perspective. And astronomers have discovered relatively youthful globular clusters in the center of a distant galaxy. Studies have revealed that globular clusters are densely packed, spherical groups of as many as 10 million ancient stars, considered some of the oldest residents of galaxies. The surprising discovery of young globular clusters, perhaps the product of a collision between two galaxies early in the life of the universe, will give astronomers insights into the way galaxies evolve.

The space shuttle will make a service stop at HST in 1994, at which time the telescope's flaws will be repaired and a new camera added.

A REVIEW OF RELATIVITY

There is little in our everyday experience of the world that will help us to understand the theories of relativity that mathematician Albert Einstein (1879–1955) formulated in the early years of the twentieth century. However, his general and special theories of relativity are well established underpinnings of much scientific activity today and merit a little review before we move on.

Einstein developed his theories in his free time while working in the Swiss patent office. His ideas, which are now a familiar old hat for scientists, have had earth-shaking consequences for us all. They led to the development of nuclear weapons, as well as nuclear power plants and particle accelerators.

An important part of Einstein's special theory of relativity is the most famous formula of the twentieth century—$E = mc^2$—which means essentially that mass, the quantity of matter in a material object, is equivalent to energy. In fact, it is a compact storehouse for an immense amount of energy: The nucleus of the atom, the basic building block of matter, is held together by the most powerful force in the universe. The amount of energy in a given amount of mass is determined by multiplying mass, measured in grams, by the square of the speed of light—30 billion centimeters per second.

Einstein's theories led to technologies that could break apart the nucleus of the atom, or force it to fuse with another nucleus in processes that release extra energy.

The first such technology was the atomic or fission bomb, developed soon after the formula $E = mc^2$ was understood. The bomb released massive destructive force by splitting apart the nuclei of atoms in radioactive material. Each fission of a nucleus produced two smaller fragments whose combined mass was less than the original nucleus. The missing mass was converted into energy.

Hydrogen bombs work by fusing hydrogen atoms together to make helium, the same process that takes place inside the sun and creates the solar energy we need to stay alive. (See THE LIFE AND DEATH OF STARS, page 26, and FUSION, page 60.) But whereas the fusion in the sun proceeds at a leisurely pace that ensures the maintenance of its energy output for billions of years, hydrogen bombs use their fuel all at once in an extravagant and destructive blast.

By contrast, nuclear power generates electricity by releasing energy in a highly controlled version of the fission that occurs in the atomic bomb. The energy released when uranium nuclei are split apart turns water into steam which drives power plant turbines and creates electricity. Nuclear power provides about 20 percent of the electricity in the United States. (See FISSION, page 64, and NUCLEAR POWER, page 66.)

It may help if we discuss in more detail the actual meaning of mass. Mass, once again, is the quantity of matter in a material object. Newton's first law of motion states that material objects continue in a state of rest or motion unless they are made to do otherwise by some outside force. A moving object will continue to move in a straight line unless some force prevents it. Mass is actually the measurement of a material object's tendency to resist this kind of change.

We often think of mass being equivalent to weight, which is something else entirely. Weight has to do with the power of gravity. A box full of lead balls feels heavy because of its gravitational attraction to Earth. On the moon, where there is very little gravity, the box and its contents would seem to weigh nothing. Nevertheless, they still would have the same amount of mass as they did on

Earth and the same amount of force would be needed to shake the box.

Newton's second law of motion states that mass stays at rest or continues to move unless some force intervenes. The state of rest or motion of the mass is proportionate to the amount of net force applied and, obviously, the greater the mass, the greater the force needed. (Net force refers to the combined force used—such as two astronauts shaking the box of lead balls—or the balance between opposing forces.) The direction in which the mass moves is the direction in which the net force is applied. If one of the astronauts pushes the box of lead balls, it moves in the direction of the push.

Newton's third law of motion describes the interaction of two masses. For every action there is always an equal and opposite reaction. The astronaut pushes the box of lead balls, but the box also pushes the astronaut. Forces always occur in pairs.

Einstein's Special Theory of Relativity expands on Newton's ideas about motion and predicts other aspects of mass and energy that are also difficult to understand because they are completely outside our experience. For example, Einstein suggested that an object's mass is affected by the speed at which it travels. The energy of its motion adds to the mass of the object so that as it approaches the speed of light, its mass increases towards infinity while its length decreases. A spaceship racing through space, therefore, would become heavier and shorter as it reached the speed of light.

Time would pass more slowly, too, although if you were riding on that spaceship, nothing would seem different to you. According to Einstein's general theory, our observations of our surroundings are relative to our own frame of reference; we carry our sense of time, speed, and dimension with us. From your perspective on the spaceship, everything would appear normal. You would not feel yourself and your spaceship contracting into a stubby bullet, nor would you sense that time was slowing down. When you returned to Earth, however, everything would have changed because time passed more quickly for everyone else. You would be able to meet your great-great-grandchildren.

No one has yet been able to make this kind of journey, but physicists have observed some of these effects in particle accelerators, as particles zoom along near the speed of light, growing fatter and more compact as they go.

Einstein also described the effect of mass on space and time. Space, according to the General Theory of Relativity, is shaped by the objects and activities in it. For example, the mass of the sun bends space in such a way that the planets travel in lines that also curve, causing them to roll around the sun in elliptical orbits.

Scientists have been able to observe the distorting effect of the sun's mass on starlight that passes by it. It is bent in such a way that stars appear to be where they really are not.

The ultimate effect of mass on the shape of space is the black hole, an object so massive that it curls space over on itself, letting nothing escape, not even light waves. No one has really proven the existence of black holes, since it is difficult to detect an object that doesn't emit any kind of radiation. But black holes may reveal their presence because of the effect they have on surrounding material, such as the blazing last gasp of stars that are falling into them. (See BLACK HOLES, page 32.)

The mass of an object also slows down the passage of time. For example, the farther you are from Earth and its gravitational attraction, the faster time passes. So if your twin lived in a space station orbiting the planet, he or she would age slightly more quickly than you, unless, of course, he or she took a trip to outer space on a ship that traveled near the speed of light, in which case time would pass more slowly and he or she would age far less than you.

THE BIG BANG

Imagine, if you can, the entire contents of the universe—space, time, and all matter—compressed into an enormously dense speck smaller than an atom. Then imagine that this speck, the tiny seed of reality, suddenly expands, swelling into a fantastically hot and homogeneous soup of matter and energy. That's the Big Bang.

How, you might ask, do we know this? Certainly no one was

around to observe the birth of the universe. However, by looking deep into space, cosmologists can see evidence that this event took place some 15 billion years ago.

As if driven by just such an explosion, the universe seems to be expanding or moving outward. As noted earlier, in the 1920s the American astronomer Edwin Hubble discovered that there are other galaxies in the universe besides our own Milky Way. Furthermore, he found that these distant galaxies are quickly moving away from us and each other, and that those farthest away seem to be moving more quickly at a speed proportional to their distance, a relationship stated in Hubble's Law. It has helped astronomers measure and make three-dimensional maps of the universe and predict the rate at which it is expanding.

Another more recent discovery has strengthened the argument for an explosive beginning of time. Instead of being absolutely cold, outer space seems to be blanketed by a uniform level of background radiation, leftovers from the boiling caldron of the early universe.

Scientists predicted the existence of background radiation long before they were technically able to detect it. They reasoned that if the universe began with an inferno, followed by expansion, there should be radiation left over, just as a dying fire radiates warmth into the cool night air around it. Then in 1964 these radio microwaves were finally detected.

This ancient energy is the most plentiful form of radiation in the universe and reveals to scientists what things were like some few hundred thousand years after the Big Bang, before gravity began to stir matter into the first stars. Infrared radiation left over from the formation of these first celestial bodies should also exist, but scientists have yet to detect it. An important step forward in the analysis of background radiation was NASA's 1989 launch of the Cosmic Background Explorer (COBE), a satellite that carries detectors designed to measure radiation without the interference of Earth's atmosphere.

The background radiation boosts the coldness of space about 3 Kelvins (K) above absolute zero. Kelvin is a temperature scale used by scientists and is named for its inventor—British physicist Lord Kelvin (1824–1907). Temperature measures the kinetic energy of atoms and molecules in matter; absolute zero, or K, is the temperature at which that activity ceases. It corresponds to −237.15° C or −459° F. The leftover energy in the universe exists in just the right amount to be consistent with the model established by the Big Bang Theory.

But the discovery in 1992 of huge wisps of gas stretching across two-thirds of the known universe may be the strongest evidence yet for the Big Bang. These wispy objects, possibly the oldest known structures in the universe, were detected and measured by NASA's COBE. Faint microwave radiation emanating from them have been rolling through space and time for billions of years.

In measuring the waves, COBE has probed events that date back to 300,000 years after the Big Bang. That's as close as scientists have come to seeing the beginning of space and time. Before then, the intense heat of the early universe prevented atoms from clinging together, so that any photons that were formed were quickly reabsorbed by this soup. Without photons, the carriers of electromagnetic radiation, there is nothing to see.

This discovery is important because it solves the mystery of how galaxies and other objects formed from what was believed to be a uniform soup of matter. The satellite's instruments found large areas in space that are slightly cooler than the rest of the universe. Scientists now think that these areas, cooler than surrounding areas by only 16 microkelvins or millionths of a degree C, are the seeds from which galaxies formed. The cool spots drew matter in upon themselves and began the process that created the clumpy universe in which we live.

By studying such footprints and retracing the steps of the expanding universe, cosmologists are describing, back to a split second after the Big Bang, what they think was the beginning of space, time, and everything else. To ask what came before that is to wander into the realm of philosophy and religion.

But where does it all end? Scientists disagree about that. The force of the initial explosion at the birth of the universe is controlled by the force of gravitational attraction between the masses of material. If this outward-rushing material goes fast enough to escape gravity, then the universe will expand forever. This phenomenon is called an open universe. In a closed universe, gravity would eventually pull the expansion to a halt and draw the mass of the universe slowly back into the speck of energy from which it all started.

A closed universe is an interesting question for scientists, but it shouldn't be a big worry for us. If everything is going to scrunch down to a speck, it won't be for billions of years. By that time our sun will have burned up most of its fuel, puffing up to a red giant that would gobble up a few planets, including ours, before the light goes out.

CLUSTER THEORY

Astronomers have mapped only a tiny portion of the visible universe. And just as early mapmakers were forced to base their assumptions about the shape of the world on incomplete information, so too have astronomers developed theories based on thin slices of evidence.

In an ambitious attempt to map the universe, today's astronomers continue the work of stargazers down through the ages, plotting on maps what they find with their telescopes. Edwin Hubble's discovery that the universe is expanding at a rate proportional to its distance from the observer has allowed researchers to make three-dimensional, rather than flat, maps of what they can see. As a result, previously undetected patterns have emerged.

Challenging the long-held notion that the universe is homogeneous, astronomers are discovering that galaxies are scattered throughout the visible universe in clumps and not spaced out in a regular way like large-scale dots on a grid. These whorls and ellipses seem to bunch together in clusters, and the clusters are part of bigger systems called superclusters. They are spread out in sheets of stars that are curved around vast voids in space. A slice of this lumpy pattern has been compared to a slice from a sponge, a honeycomb, or a sink full of bubbles. The voids seem to be common throughout the portions of the universe that have been surveyed.

Some are larger than others, as though some of the bubbles or cells of emptiness had merged.

Some galaxies exist in the voids, but they are much fewer and less dense. The existence of other invisible matter in the voids is inferred by the gravitational behavior of visible stars and galaxies. If scientists knew what was there, they might have a better idea as to how the pattern developed.

Gravity is responsible for much of the structure in the universe, but that alone is not enough to explain the presence of these massive blank spots in space—unless some invisible matter is out there exerting its influence. Researchers believe there surely must be some such matter filling the space within and around the visible galaxies, and if so, perhaps it fills the voids as well. Then the material that evolved into galaxies may have clumped together where the dark matter is especially dense.

The voids may have been created by tremendous explosions in the early ages of the universe, although no one knows what might have caused them. Germinal material, mainly hydrogen, may have been blown out into shells, which eventually coalesced into the clumpy layers of galaxies that are visible today.

As telescopes and other astronomers' tools become more sophisticated, scientists will be able to look deeper and deeper into outer space and its past. Maybe this mystery and all the other puzzles about our universe will eventually be explained.

COSMIC STRINGS

Like dew drops on a spider web, the structures of the universe might have developed on a network of cosmic strings. According to this hypothesis, great strings of very thin but immensely dense cosmic plasma (to the tune of 20 quadrillion metric tons per inch) were formed as the universe cooled just after the Big Bang. These tangled and sometimes looped strings may have wriggled about across the universe, attracting the scattered material from the Big Bang into clumps or clusters that ultimately became galaxies.

Cosmic strings may have disappeared long ago, but there is some evidence that they still exist. The appearance of mirror-image galaxies, seemingly two galaxies where there is really only one, suggests to some scientists the presence of a great magnetic influence that bends the light of a distant galaxy in a manner sufficient to make mirages in which the galaxy appears as multiples of itself.

DARK MATTER

There are other explanations for the clumpiness of a universe that is thought to have started out smooth. It may be full of strange shapes or particles that can't be seen or detected but still affect the behavior of the visible matter with their presence.

For centuries, astronomers have assumed that the luminous bodies visible in the sky accounted for most of the mass of the universe. Only recently have they realized that the matter that is visible may constitute as little as 1 percent of the mass. The rest, if it exists, would not radiate any energy that is detectable. Scientists suspect this dark matter is there, though, because of its gravitational effect upon the visible stars and galaxies.

Astronomers have assumed from their observations that the greatest mass of a galaxy exists at its center and have built their understanding of the behavior of these apparently center-heavy structures on the model of our own solar system: The greatest mass resides in the sun, and the planets closest to it orbit at a greater velocity than those that are farther away and less affected by the sun's gravitational pull.

There appear to be several kinds of galaxies in the universe. Spiral galaxies are flat discs of light, with outstretched arms in swirling curves around a more densely illuminated bulge in the center. Others are elliptical, and many are small irregular blobs.

Since velocity is a measure of gravitational pull, those stars on the outer edges of galaxies should be orbiting more slowly, but this is not the case. The velocities of stars on the outer edges were found to be the same as those near the center. Astronomers have thus concluded that great mass must exist where there is little light.

Furthermore, these galaxies congregate in enormous clusters, seemingly held together by their gravitational attraction to each other, even though there isn't enough visible mass to account for all this gravitational power.

The only conclusion, if we wish to keep our Newtonian understanding of gravity, is that there is more out there than meets the eye—or any other observational device we have. But what is it? It could be particles left over from the Big Bang, intergalactic rocks or planets and lots of them, or the various undetected evolutions of stars, including white dwarfs and black holes. But truth be told, no one knows.

THE LIFE AND DEATH OF STARS

Stars are born from massive clouds of gas, mostly hydrogen. Gravity forces them to collapse under their own weight, and the resulting density and heat create an atomic traffic jam in which hydrogen atoms collide with enough force to fuse into helium, the gas that makes balloons float above the heads of happy children. The energy released by that fusion lights the star's candle, so to speak. Meanwhile, the outward pressure of the released energy pushes against the force of gravity, which is still trying to pull everything in toward the center.

After the hydrogen in the star's core has been used up, gravity begins to overpower the star's ability to emit energy. The heat created by this compression forces the helium to fuse into carbon atoms and the resulting energy release continues the fight against gravity. Carbon then fuses into oxygen and neon, neon into silicon, and silicon into iron, which sounds the death knell for the star. As each new fuel is exhausted, another takes its place, until the star resembles a many-layered onion of elements.

Of course, each star eventually loses its battle against the force of gravity. Its size and proximity to other stars determine how long it will survive and how its life will end. A smaller star is less luminous, but the trade-off is a much longer life.

Our sun will burn up its fuel in about 4 or 5 billion years. When it has passed its prime, it will swell from a yellow-white light into a huge red beacon called a red giant. Then it will cool and shed its outer coat of elements into the soup from which new stars are made, leaving behind a dim white-hot core that cools steadily. This white dwarf will be able to maintain an equilibrium between its energy and gravity for a very long time. (See DWARFS AND RED GIANTS, below.)

Giant stars many times the size of the sun seem to have a more dramatic fate, however. Because they are big, they have to burn a lot of energy to counteract the tremendous force of gravity their size demands. Ironically, the bigger the star, the sooner it dies; smaller stars like our sun burn more frugally and live much longer.

What is the fate of these powerful giants? After they've exhausted their fuel, the violent collapse caused by gravity makes them explode. This bright flash of energy, called a supernova, creates and releases the heavier elements that will form new generations of stars. Each generation has higher concentrations of the elements released after stars explode. Our sun is probably a third generation star born in a rich cloud of material that formed our planet and made life possible.

In fact, scientists believe that stars, using hydrogen and helium created by the Big Bang, are factories that produce the building-block elements of the universe. The hemoglobin in our blood, the calcium in our bones, the stones beneath our feet are all inherited from the stars.

DWARFS AND RED GIANTS

The Milky Way galaxy is brimming with dwarf stars. But dwarfs are not really small, as their name might suggest. The name classifies a star according to its temperature and brightness.

Our sun is a typical yellow-white dwarf that sits near the middle of the scale of luminosity. Ninety percent of the stars near us are dwarfs, which means they are middle-aged, still burning hydrogen in their cores with plenty to spare. Our sun, for example, has been steadily burning hydrogen for 5 billion years and probably has enough fuel to last another 5 billion.

Dwarfs that are brighter than the sun are also more massive and burn with a blue-white light. While they may have more hydrogen fuel to burn, they must burn it furiously in order to maintain their stability. They don't last long, as stars go. They use up their fuel stores in hundreds of millions of years.

At the other end of the brightness/temperature scale are the orange and red dwarfs, which are much dimmer but could hold out for a trillion years.

When a dwarf like the sun runs out of hydrogen, it becomes less stable, contracting and expanding as it burns up the last shreds of fuel in its core. As the outer layers expand, adding surface area, the star cools and temporarily becomes a "red giant," burning everything it can to survive. Deep in its belly, it forges helium into carbon. But the outer layers continue to cool and disperse, leaving a hot, dense core called a white dwarf. At this point, gravity has forced a mass similar to the sun into a pretty tight fit, an area as small as the earth.

As the dwarf loses heat, it manages to maintain its equilibrium with pressure exerted when the highly compressed and agitated electrons in the core repel each other. So the center holds, but the white dwarf eventually becomes a cold, dark, dead thing called a black dwarf. Or so scientists think. No one knows if such a thing really exists or has even had time to evolve.

Another hypothetical member of the dwarf family is the brown dwarf, a ghostly presence that is more massive than any known planet, but not massive enough—at least 8 percent the mass of the sun—to trigger the fusion that would make it a star. These dim bulbs may orbit around greater stars in the firmament or actually ply their oars in a solitary journey through space. Their presence has been revealed only by the tremor they create in the motion of more visible stars.

Recently, however, astronomers at the University of Minnesota, using a high-speed scanner, examined photographs taken in 1955, 1962, and 1975 of the Hyades star cluster in the constellation Taurus with the telescope at California's Mount Palomar Observatory. By comparing these old photos of dim specks with each other and by examining new infrared data collected with a telescope at Mauna Kea in Hawaii, the Minnesota astronomers concluded that twelve huge dim stars exist in the Hyades cluster that could very well be brown dwarfs. If they are real, these barely visible bodies may account for some of the missing mass that confounds science.

SUPERNOVAE

Chinese astronomers, scouring the heavens for signs and meaning some nine hundred years ago, found what they thought was a new star. But they were wrong. What they witnessed was the violent death of an old star. Some distant giant was losing a battle with its own gravity, exploding with the energy and force of a hundred billion suns. Chinese records report that the phenomenon was visible during daylight for 23 days. The Chinese referred to it as a "guest star"; modern astronomers call such an explosion a supernova.

Although the Chinese and Japanese documented the 1054 supernova, until recently there was no record found of the event in the Western Hemisphere. Now, two astronomers at the University of Texas have found what they believe is a depiction of the supernova on a piece of pottery dating back to the period.

In 1987 a known star exploded in the Large Magellanic Cloud, near our own galaxy. It was the first supernova since the seventeenth century to be visible to the naked eye. This important event has allowed astronomers to compare the spectrum of the star before and after it exploded and learn more about supernovae.

Huge stars, at least eight times the mass of the sun, extinguish their fires in supernovae that occur about once every hundred years in the average galaxy. When these stars have used all their fuel, gravity drives the remaining material down into the dead iron heart of the star. An explosion just outside the core compresses it further and blows the rest of the star out into space.

The debris from the explosion rolls away from the star's core in an expanding shell of frantic electrons, which smash into the gases that drift between the stars. The shock of the explosion may trigger the formation of new stars, while the friction of colliding particles creates a wispy glow that will last for thousands of years. The Crab Nebula in the constellation Taurus is thought to be the remains of the "guest star" that Chinese astronomers observed in 1054.

The Crab Nebula and some other supernova remnants have within them a pulsating source of radiation that scientists believe to be a neutron star. The gravitational plunge of material during the explosion shears apart the atoms in the star's core, fusing electrons and protons into a tightly wound glob of neutrons, which spins at a speed that would rattle apart any other star.

A supernova may end as a black hole, the dark grave of a dead giant. The most massive stars collapse to a density so great that gravity allows nothing, not even light, to escape.

But not all supernovae begin as giants. Apparently, some occur in pairs of smallish stars that are locked in a close dance around each other called a binary system. One of the stars is a white dwarf. Perhaps it greedily unwinds material from its ordinary sister star, or maybe the ordinary star carelessly drops its hydrogen scarves onto the dwarf. In either case, when the dwarf acquires more mass than it can carry, the balance between the outward thrust of energy and the inward pull of gravity is lost. The heat generated by its sudden collapse lights one last atomic blast and the dwarf explodes. Its partner is blown out of the way and becomes a runaway star.

NEUTRON STARS AND PULSARS

Though we talk about the death of stars, it is apparent to physicists and astronomers that old stars don't die; they just compress. At least some part of them does, anyway. After they explode or slowly bleed their outer layers into interstellar space, they find other roles to play in the business of the universe.

When it runs out of fuel, a giant star can no longer exert pressure to counterbalance the inward pull of gravity. The star's core collapses and causes a tremendous explosion, shooting its payload of manufactured elements out into space. Depending upon the mass of the star, this explosion or supernova can result in either a neutron star or a black hole.

Neutron stars are not just the collapsed corpses of exploded giants. They appear to be tiny but powerful beacons in the dark outposts of distant galaxies. In a neutron star, the force of the explosion crunches the core of the star into a strange dense ball that is only a few miles in diameter. This tight little wad of stardust is so dense that one cubic inch could weigh 50 billion tons. The pressure is such that protons, neutrons, and electrons crush together to form a compact ball of neutrons that scientists compare to a giant nucleus. The natural repulsion among neutrons is sufficient to withstand any further crushing by gravity. If the neutron star's mass were equivalent to the mass of three or more suns, it would collapse into a black hole.

The interior of a neutron star is so dense that its contents are superfluid, which means they flow without friction. Other strange liquids and exotic particles may also be locked within the immensely strong crust of iron atoms knitted together by the powerful magnetic field on the surface of the star.

Rapidly spinning neutron stars that emit strong pulses of energy are called pulsars. Just as figure skaters increase the rotational speed of a spin by pulling their arms and legs closer to the axis of that spin, the rotating neutron star spins faster and faster as its mass is squeezed in by gravity. Meanwhile, the powerful magnetic lines coursing over its surface become billions of times more concentrated, so that as the neutron star spins, it radiates energy from its magnetic poles, blinking up to several hundred times per second. Earth could set its watch by the regular pulses of energy emitted by these spinning relics of old exploded stars.

Sometimes pulsars seem to be part of binary systems. Some of the extremely fast pulsars may actually be exhausted neutron stars that have been kicked into a faster spin and more frequent pulses by the momentum of material falling from red giant companions. But of the four hundred neutron stars that have been identified in our galaxy, we know of only four pairs of orbiting neutron stars.

These oddball celestial objects were predicted over sixty years ago when the neutron was discovered. But it was not until 1967 that evidence of their existence was found. Since then, their peculiar ways have been the object of intense interest among scientists.

BLACK HOLES

Black holes could be called the ultimate prison. They are objects formed by the collapse of gas clouds, stars, or groups of stars around which gravity is so intense that nothing—not matter, space, time, or light—can escape.

It's possible that there are black holes left over from the early days of the universe when the stringy, globby primordial soup became so dense in places that gravity caused some areas to collapse upon themselves. Such density hiding in the dark folds of space may account for some of the dark matter or missing mass that scientists can't see but are sure exists.

Black holes might also be the immensely dense relics of dead stars. When a star several times the mass of the sun has used up its fuel, it has no outward pressure with which to fight the effect of gravity. The violent inward crush of material would cause the star to explode, driving vast amounts of material out into space, while pressing the remaining core of the star, still massive, into a much more compact package. It would be like compressing the mass of the earth into an object that is one-third of an inch in diameter. The gravitational power would be strong enough to bend back any beams of light trying to leave it. That's why it's called a black hole. Since nothing we know of travels faster than light, nothing else could escape from the object, either.

If light can't escape from a black hole, it is fair to ask how scientists know that such things exist. They have observed that binary star systems—pairs of stars that revolve around a common center of gravity and may account for as many as half of all stars—sometimes include bodies that emit strong X-ray signals and are too massive to be dim dwarfs or neutron stars. The X-ray signals are probably generated by friction in the disk of material as it swirls into the black hole from its partner star. In this situation the black hole might even appear to glow, when, in fact, it is the material falling into the hole that is emitting light.

Massive black holes may be responsible for quasars, the most brilliant objects in the universe and remnants of the violent beginnings of the universe. It is thought that these black holes were created by the collision of stars deep in the centers of galaxies. The friction caused by stars and other material falling into the hole creates heat and light that is visible over tremendous stretches of space and time.

What may be a massive black hole was detected at the center of the galaxy M87 some 50 million light years away in the constellation Virgo. Using the Hubble Space Telescope, scientists were able to collect new evidence for the theory that black holes are the power behind the brilliance of quasars. They saw an enormous jet of flaming gases emanating from the heart of the galaxy, similar to observations made by astronomers earlier in this century. Using HST they were able to detect a density five hundred times greater than earlier ground-based telescopes had revealed. Whatever it is, the object at the heart of M87 contains as much matter as 2.6 billion suns, and the resulting gravitational force seems to be swallowing stars.

Evidence for the existence of black holes continues to accumulate, yet no one has seen or proven the existence of these strange things. The idea began as mathematical conjecture based on theories of physics that were created before the invention of earth-orbiting space telescopes. Now astronomers are getting closer and closer to understanding the mysteries that the theories predicted.

QUASARS

Whole chunks of galaxies may suffer the same fate as dying stars, creating giant black holes and a flash of explosive energy that is called a quasar, short for "quasi-stellar radio source."

Quasars appear to be billions of light years away and are the most brilliant things in the universe, some as luminous as a trillion suns. They look like stars, but upon examining the spectrum of wavelengths emanating from these so-called stars, astronomers were puzzled to find elements that weren't starlike at all. About 10 percent emit strong radio signals in addition to infrared, X-ray, and gamma-ray frequencies. In fact, scientists have recently identified four quasars as the source of the ultrahigh-energy cosmic rays that seem to flood the universe. These quasars are pouring out huge rivers of gamma rays. Stars, on the other hand, emit most of their energy as visible light.

These starlike sources of radio energy also exhibit a shocking red shift, indicating that these quasars are very far away and receding at an astonishing speed. The fact that we see them means they must be very luminous indeed.

Furthermore, all that energy seems to emanate from a source that is very small relative to the energy it releases. A quasar's light can change in a few days, leading astronomers to believe that its radius can't be greater than the distance light can travel during that time. That's because light from the back and side edges of a luminous object must travel to the leading edge of the object before changes in total brightness are detectable from our distant vantage point. Many of these immensely distant objects may be just one light-day across (the time it takes light to travel in a day), about the size of our solar system, and yet they shine a thousand times brighter than an entire galaxy.

The dark secret at the heart of these bright objects may be a monster of a black hole, with the mass of a hundred million suns, swallowing a star every year as well as great volumes of gas and dust. The light emitted by quasars probably comes from the edge of the hole where these materials swirl in a brilliant skirt of colliding particles. Nearer the edge of the hole, the particles move faster, some 3,000 miles per second, and the friction from their dance is released as heat and light. This accretion disk, as it's called, is a strong source of ultraviolet radiation. Perpendicular to the disk of matter falling into the hole are jets of particles, another source of radio signals, being spit from the quasar at the speed of light.

Some astronomers believe that quasars are the flashy centers of young galaxies formed during the infancy of the universe, and that such activity may be a natural stage in galaxy development. By viewing them we may understand better what lies hidden by gas and dust in the nucleus of spinning disks of stars like our own Milky Way.

GLOBULAR CLUSTERS

Globular clusters are rather like senior-citizen communities for stars, although there is no evidence that they moved there to get away from the weather and the rat race in some younger galaxy. It's not clear how they got there, but scientists think these beautiful spheres of ancient stars, like groups of wise and experienced grandparents, can teach us much about the history of the universe.

In our galaxy alone, scientists have identified some 150 globular clusters that may be 15 billion years old. They are probably older than any other community of stars in the universe, each containing as many as a million ancient stars crowded together into much too small a space. The crowding in these communities creates what might at first appear to be an unfortunate undercurrent of violence. The stars can't get around without bumping into each other, but rather than destroying the community by causing it to explode into space or sink dismally into a black hole, the situation may actually serve as the globular cluster's fountain of youth.

Beaming out from these clusters of stars are strapping, young, hot-blue stars called blue stragglers. In a community as old as this, the bigger stars should have exhausted their fuel long before. Globular clusters usually lack the pools of gas and dust that give birth to new stars. Intense magnetic fields created by collapsed stars apparently shove this material out of the star system, preventing the advent of disruptive infants. So why the blue stragglers?

Scientists think that the stars in globular clusters rejuvenate themselves by bashing into one another and coalescing into massive new stars. This activity stirs up the juices in the outer layers of star pairs, drawing hydrogen into the core of the newly created star, providing new fuel to burn—relighting the old candle, as it were.

Furthermore, there is some mechanism at work that shoots stars from the core of globular clusters like stellar cannonballs. Scientists believe that the crowded conditions create a favorable environment for the formation of binary systems, with gravity allowing stars to draw one another into brief orbital relationships.

This notion is supported by the discovery of pulsars in globular clusters. Pulsars are old stars that have collapsed into dense relics called neutron stars. As they collapse, neutron stars rotate faster. As it shrinks, the star still retains its magnetic field, which becomes powerfully concentrated. The combination of spin and magnetic energy causes the neutral star to emit a regular pattern of radio signals, blinking like a great cosmic lighthouse. Eventually pulsars run out of spin and settle down to a quiet retirement, unless they somehow find a way to jump-start their hearts by increasing their mass.

Stars as old as those in globular systems would normally be enjoying a quiet retirement, but the crowding near the center apparently allows pulsars to hook up with other old stars and pick up speed by peeling off some of a partner's mass and angular momentum. These whirling partners act, in turn, like giant eggbeaters, slinging material from the core of the cluster out to the periphery, keeping the center from collapsing into a black hole, something that scientists believe happens in most galaxies.

Globular clusters dance around the outer edges of our galaxy and seldom meddle in its business except to pass through its disk a couple of times during their own orbits. By studying globular clusters, astronomers at the turn of the century learned that our solar system is not at the center of the Milky Way but about two-thirds of the way out on the leading edge of one of the great spiral arms. Now with the help of the Hubble Space Telescope, scientists are able to study more closely the internal workings of globular clusters and find out more about the evolution of galaxies and the interaction of stars when they are crowded together at the center of huge systems.

The pulsars within globulars are also tools for measuring the contents of the galaxy. Because the regular blink of the pulsar is changed by the presence of interstellar gas, any alteration in the pulse helps scientists measure the density of what they believe is a thick layer of ionized gas blanketing the solar system. The gravitational tug of massive structures also affects the pulsar's blinker, allowing researchers to map more accurately the distribution of mass in the galaxy.

GRAVITATIONAL LENSES

Sometimes the mass of a star or galaxy distorts the image of a more distant galaxy or quasar, magnifying it or even splitting it into two or more identical images. In effect, the power of gravity is bending the light, a phenomenon called gravitational lensing that Albert Einstein predicted in 1919. More than sixty years later, astronomers identified the first example of gravitational lensing.

To visualize how it works, try the simple lesson of standing a pencil in a clear glass of water. The bottom half is magnified and the pencil looks as though it's cut in half. Perhaps you have also seen a mirage, in which an image of some distant place shimmers eerily above the horizon. On Earth, layers of cold and hot air bend light to create watery images that float near the horizon. Sometimes distant images of real things are projected into our line of sight by the refraction of light.

In the case of gravitational lenses, it is the gravitational energy surrounding massive objects that does the job, causing the light from a distant quasar to split, with some of it taking what seems to be a longer road. One way astronomers were able to confirm this phenomenon was by comparing patterns in separate streams of light coming from what appeared to be twin quasars. A quasar's bright light can vary in intensity over a short time, somewhat like a flickering candle, but if the variation patterns and other factors match, it's reasonable to think that the light streams came from the same object.

Sometimes gravitational lenses fool astronomers into thinking that a bright object is actually part of a nearby galaxy. In a variation called a microlense, the gravitational power of an individual star passing through our line of sight bends the light from a distant object and projects its image onto the nearby galaxy. The individual star can also brighten the distant light stream and make it seem to flicker.

Because gravitational lenses and microlenses are actually extremely powerful telescopes, the magnification of galaxies and quasars will allow scientists to study objects in new ways. For example, astronomers hope to determine the masses of these intervening stars by observing microlenses. If a galaxy of such "magnifying glasses" is watched over the years, it should be possible to measure the stars, including those not otherwise visible, like neutron stars, brown dwarfs, and black holes. Then it may be easier to identify the missing mass or dark matter that seems to make up over 90 percent of the universe.

Another useful tool for astronomers is the time delay between the arrivals of the separate streams of light or images. By calculating the delay between these images, scientists have a second chance to observe interesting phenomena emanating from quasars. What's more, delays measured in years also give astronomers time to prepare for a second chance to view and study the image or light stream.

Gravitational lenses also create rings that are detectable with radio telescopes. All the matter in a nearer galaxy refracts the radio waves emitted by an object billions of light years behind it, pushing this energy out into enormous halos. These Einstein rings, as they are called, may help astronomers map the distribution of visible and invisible matter in the universe.

Arclets are another type of ring that forms when massive clusters of galaxies refract the light of more distant galaxies. In the foreground is the red light of the nearby galaxy. Peeking through are little arcs of light that are the distorted blue-light lines of faraway younger galaxies. They, too, may prove useful as mapping tools.

THE SOLAR SYSTEM

Scientists don't pretend to know *why* the solar system was formed, but they think they understand *how* it happened. And, sad to say, the known planets in the solar system—Mercury, Venus, Earth, Mars, Jupiter, Saturn, Uranus, Neptune and Pluto (in addition to over fifty moons and an untold number of asteroids, comets, and other chunks of debris)—may be the incidental leftovers of a much grander purpose, the formation of a star.

The planets were formed when, under the influence of gravity, a slowly rotating cloud of gas and dust began to shrink into a fledgling star. Matter trailed out around the center in a disk, and gravity caused the motley parade to form clumps that eventually became planets. It's not clear whether the planets took shape as the sun did, collapsing from clumps of gas, or whether they started out as particles that gathered up matter as they rolled along. In either case, it was probably a messy process with many collisions. Today's solar system is still littered with aimlessly wandering debris.

The four planets closest to the sun, including ours, are small and rocky. Mercury is about half the size of Earth and, like all the planets but ours, uninhabitable by life as we know it. Its elliptical orbit carries it as close as 28 million miles and as far as 43 million miles from the sun. As Mercury orbits the sun and slowly turns on its axis, it is always very hot (about 800°F) on one side and bitterly cold on the other.

Venus, which is about the same size as Earth and often referred to as its twin, is not as friendly. Shrouded by dense clouds, it is an example of what the greenhouse effect can really do. The clouds trap the heat or infrared radiation of the sun, raising the temperature to 900° F, hot enough to melt lead. Unmanned Soviet and American spacecraft have probed this planet, which is 67 million miles

Position and Relative Size of the Planets

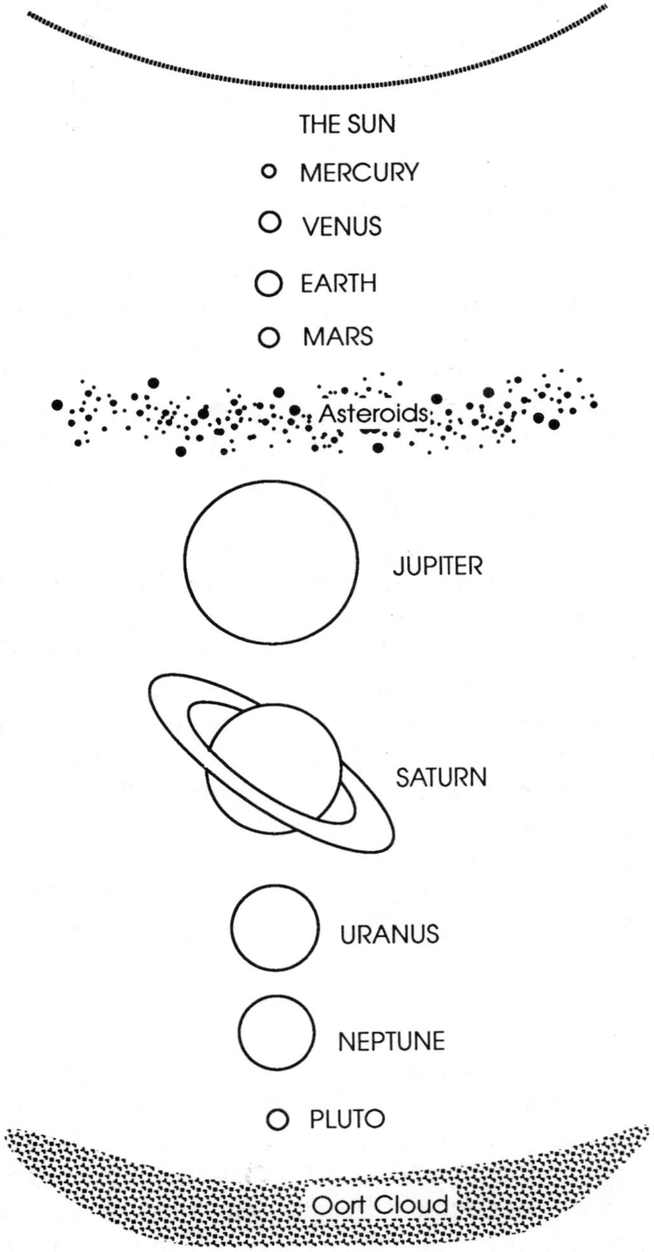

THE SUN

MERCURY

VENUS

EARTH

MARS

Asteroids

JUPITER

SATURN

URANUS

NEPTUNE

PLUTO

Oort Cloud

PLANETARY DATA

	Diameter in miles	Surface gravity (earth = 1)	Time for one rotation (in earth time)	Time for one revolution (in earth time)	Average distance from sun In millions of miles	In AU (earth = 1)
Mercury	3,030	0.4	59 d	88 d	35.97	0.387
Venus	7,550	0.89	244 d	225 d	67.24	0.723
Earth	7,944	1.0	23 h 56 m	365.26 d	92.96	1.0
Mars	4,200	0.4	24 h 37 m	687 d	141.6	1.5237
Jupiter	88,700	2.54	9 h 50 m	11.8 y	483.6	5.2028
Saturn	74,880	1.07	10 h 40 m	29.5 y	886.7	9.5388
Uranus	32,200	0.8	12–24 h	84 y	1,784.1	19.1914
Neptune	30,750	1.2	15–20 h	165 y	2,794.5	30.0611
Pluto	2,000	?	6 d 9 h	248 y	3,674.6	39.5294

m = minutes, h = hours, d = days, y = years, AU: Astronomical unit.

from the sun, finding enormous mountain ranges, evidence of volcanic activity, and clouds made of carbon dioxide and sulfuric acid.

Earth was lucky to occupy an orbit 93 million miles from the sun, just the right distance for life to take root. The axis of rotation tilts our planet 66.5° against its orbital plane, a convenience that gives us seasons.

Earth's size and location allowed it to condense from gas and star dust into a rock ball that is still hot at the core. Gases escaped from this boiling center and created a primitive atmosphere that was full of water vapor, nitrogen, and carbon dioxide. Because there was more water than the atmosphere could hold, the oceans formed; they began to drive a weather system that cycled water in and out of the atmosphere.

Rain washed carbon dioxide out of the sky and onto Earth where it reacted with minerals and became trapped in the rocks. Some of this CO_2 is returned to the atmosphere in volcanoes that erupt at the edges of plates, the segmented and mobile crust of Earth. (See PLATE TECTONICS in Part III, page 131.) Nitrogen became the most plentiful gas in the thin shell of atmosphere that allowed life to begin in the oceans. Luckily for us, conditions evolved that were right for photosynthetic plants, which pumped oxygen into the atmosphere.

Of the four innermost planets in the solar system, Earth is the only one with a relatively large moon. Rocks carried home by Apollo astronauts revealed that the composition of our moon is different from that of Earth, suggesting that it may have originated somewhere else. One theory suggests that a planet the size of Mars careened into Earth, melting the rocky outer crust and sending its iron core deep inside. The intense heat cooked off the volatile substances, which would explain why the moon has no water, chlorine, or potassium. A plume of debris settled into a ring around Earth, coagulating into a large moon, which is more like a companion planet because of its relative size.

Mars, the next planet out and 141 million miles from the sun, wasn't so lucky, even though scientists think it started out with the same sort of primitive atmosphere. Half as big as Earth and a bit too far from the sun, Mars couldn't hang onto the internal heat created by pressure during its formation. It became less and less volcanic and, consequently, less able to pump enough nitrogen and CO_2 into the atmosphere. The greenhouse effect that warmed its surface evaporated, along with much of the water. Because Mars is small, its weak gravity allowed most of its nitrogen to escape and be broken down by the ultraviolet light of the sun.

Now Mars is a cold, dead planet with an average temperature of –60° F and a thin atmosphere that is mainly carbon dioxide. Any remaining CO_2 is probably locked in the polar icecaps along with what little water may exist on the planet. Nevertheless, Mars is the only other planet in the solar system where humans might be able to survive. Scientists once hoped to find life there, but when the Viking 1 and 2 orbiters landed on Mars, they found none.

Between Mars and Jupiter lies the asteroid belt, full of a million or more boulders that range in size from a few yards to some 80 miles in diameter. Seven thousand of these fragments or planetismals have been observed and tracked, and three thousand have been named. (The Soviets named two after Herbert Hoover!)

Fragments of these space boulders get chipped off in collisions and fall to Earth, where scientists examine their composition to learn about conditions during the birth of the solar system. Carbonacious chondrites, carbon-rich fragments chipped from the outer edges of asteroids, contain crystals and other particles that are older than the solar system.

Sometimes a huge chunk of rock and iron will get bounced out of its orbit in the asteroid belt and head straight for Earth. An 820-foot asteroid came within 500,000 miles of us in 1989. It was a very close call, and no one knew about it until afterward, although astronomers try to keep an eye on such things. A collision with a sizeable asteroid would have catastrophic consequences, even worse than a nuclear war. In fact, some scientists blame the extinction of the dinosaurs on impacts from either asteroids or comets, which would have thrown up huge clouds of dust and debris, obscuring all light.

The contents of the asteroid belt might be the remainders of a planet destroyed by an even worse collision early in the life of the solar system. It might also be material that could never develop into a planet because of the gravitational influence of nearby giant Jupiter.

Of all the planets, Jupiter is the biggest. In fact, it's twice as big as all the rest put together and three hundred times the size of Earth, although much less dense. That's because it and the next three big planets—Saturn, Uranus and Neptune—are mainly layers of gaseous, solid and liquid hydrogen, helium, methane, water, and ammonia wrapped around rocky cores. Jupiter is 483 million miles from the sun and is racked by violent storms and lightning. Thirteen moons orbit this giant planet.

Scientists think the gases that blanket Jupiter and its neighbors were blown out there by powerful solar winds, leaving behind mostly rock as the building material for the first four terrestrial or earthlike planets in the system. The Jovian planets, as the big four are called, all have ring systems, sometimes barely visible, and are orbited by multiple moons. The Voyager space probes, launched in 1977, have given scientists close-up views of the four gaseous giants, revealing surprising details—previously undiscovered moons and faint rings of ice, rock, and methane.

Saturn, with its spectacular rings, appears to have ten major satellites or moons and is the second largest of the planets. It is 887 million miles from the sun and has shiny rings that are filled with ice-covered rocks, possibly the remnants of a broken moon that strayed too close to the planet. The atmosphere, like Jupiter's, is primarily hydrogen and produces violent storms.

Uranus and Neptune are big chunks of water and molten rock, with atmospheres made of hydrogen, helium, and methane. Uranus, which is 1.8 billion miles from the sun, is tipped over on its side so that the line of its axis points at the sun as the planet rolls sideways through its orbit. It is girdled by mysterious black rings. Neptune, the smaller of the two, is 2.8 billion miles away from the center of the solar system. Unlike its more sedate twin, Neptune is racked by violent storms and streaked with bright white clouds. The methane in the planet's outer atmosphere absorbs red light, giving it an aqua-blue color.

It took Voyager 2 twelve years to reach this little-known planet where the sunlight is one-thousandth as strong as it is on Earth. The space probe found six ragged new moons, observed giant storms and brutal winds sweeping Neptune's surface, and photographed the planet's four faint rings for the first time. The space probe sent back photographs of Triton, Neptune's largest moon, which was encrusted with nitrogen-and-methane polar icecaps. Slightly smaller than our moon, Triton had active geysers and showed signs of having once been a much warmer place. This moon actually may have been an independent planet that was captured by Neptune's gravity.

The sun is merely a bright and distant star to Pluto, which is billions of miles from its warmth. This ninth and outermost known planet in our solar system is half the size of Earth and very cold, with a surface temperature of about −370°F. Pluto and its recently discovered moon are probably made of ice and rocks, and scientists have wondered why it is so different from the giant gaseous planets with which it shares the outer limits of the solar system. They have speculated that Pluto might be a captured comet or a former satellite of Neptune.

Another player in the business of the solar system is the spherical cloud of comets that apparently surrounds the solar system. The Oort cloud, named for twentieth century Dutch astronomer Jan Oort (1900–) who suggested its existence in 1950, is billions of miles from the sun and may contain trillions of comets that orbit the sun in broad loops. These balls of ice, dust, and carbon are thought to be leftovers from the formation of the solar system. Passing stars and the tug of Jupiter can throw comets into orbit close to the sun and that's when we can see them. The heat and solar wind cause visible tails of water vapor and dust to trail for millions of miles.

Some scientists think that the great periodic extinctions suffered on Earth are caused by comets that are disturbed from the Oort cloud by a mysterious companion to the sun, a small star that some have called Nemesis. Its gallactic orbit brings it near the sun and through the Oort cloud every 26 million years, knocking comets into outer space and also into crazy collision courses with planets like ours.

The notion of such a star is controversial, but it has opened the eyes of scientists who thought that other planetary systems harboring intelligent life could exist only around more stable single stars. What if the evolution of intelligent life is dependent upon periodic extinctions brought on by the disturbances of a companion star?

The search for planets orbiting other stars has been an important goal of modern astronomy, but until recently, there had been little evidence of other solar systems like ours in the galaxy. Then in 1991, astronomers finally found evidence that two or more planets are orbiting a recently discovered pulsar—a dense, dead star, over a thousand light years away from Earth in another part of the Milky Way. The neighborhood of a pulsar is an unlikely place to find planets, scientists say, so solar systems of various types may be more common than anyone has thought. It's just a matter of looking in all the right places.

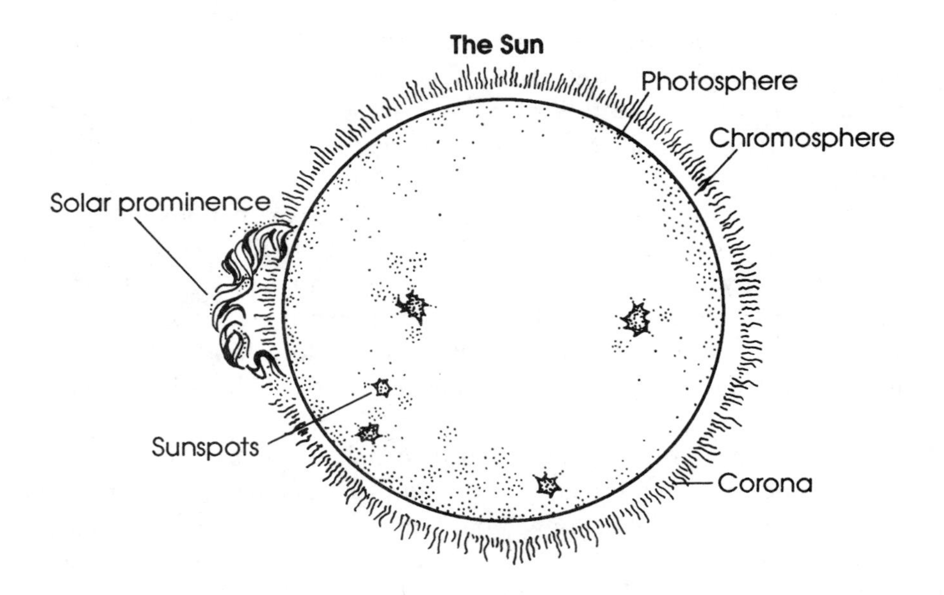

SUNSPOTS

In 1610, Galileo, using his telescope, observed dark spots on the surface of the sun. He thought at first that they were clouds, but scientists learned later that the spots—dark only in relation to the overall brightness of the sun's surface—are actually cooler areas of the photosphere, or visible surface of the sun.

The sun is a ball of gas whose core is denser than ordinary rock and whose outer layers are thinner than the air we breath. Even though it is not the brightest star in our galaxy, its interior temperature, where fusion occurs, is measured in the millions of degrees. This energy is transmitted outward through the outer layers by X-rays and ultraviolet light.

The photosphere, the outer 20 percent of the sun, is a bubbling layer through which convection carries hot gases up to the surface, where the temperature is about 5,500°C, and cooler gases back down. The intense bubbling heat of the gases in the photosphere give it an opaque and bumpy appearance, dotted by sunspots and fiery plumes of gas. From the surface, the sun's energy radiates into space, bathing the solar system with varying degrees of warmth and light.

Observers have known since the nineteenth century that sunspots occur in cycles that reach their peak every eleven years. At the low ebb of the cycle the sun's surface appears to be a golden blank, and at the height of a cycle there can be as many as a hundred spots a day, moving across the surface from left to right over a period of about fourteen days.

The sun's energy output seems to waver by small amounts during sunspot cycles, and researchers have sifted through centuries of astronomical records to find correlations between sunspot activity and weather conditions. During a seventy-year period encompassing parts of the seventeenth and eighteenth centuries, sunspot activity was infrequent. Some scientists have linked this period to the Little Ice Age in Europe, which, ironically, included the reign of France's "Sun King," Louis XIV.

A link between sunspots and the energy output or solar constant of the sun was not confirmed, however, until the 1980s when NASA's Solar Maximum Mission Satellite (Solar Max) was launched. Information gathered by Solar Max and NASA's Nimbus 7 weather satellite showed that sunspot cycles seem to be related to a network of greater magnetic activity that intensifies as the sunspot cycle reaches its peak.

The sun has a dipole magnetic field that is similar to, but much stronger than, Earth's. Sunspots, some of which are three times the diameter of Earth, are surrounded by magnetic fields that can be more than a thousand times stronger than elsewhere on the sun's surface.

Also coinciding with the appearance of sunspots is the emergence of even larger areas that appear brighter than the surrounding surface. This patchwork of bright and dark spots—embedded in powerful magnetic fields and superimposed on the blazing sun—seems to cause the waxing and waning of the solar constant.

Investigations of sunspot activity have played a part in the development of radio astronomy, the study of radio waves emitted by celestial objects. An incident during World War II served as a catalyst. When the British army's radar was jammed so effectively that two of their own battleships coasted undetected through the English Channel, physicist James Stanley Hay was assigned to figure out how it was done. He discovered that the interference came not from Nazi installations on the coast of France but from the sun. A large sunspot had recently occurred, and the accompanying strong magnetic field interacted with electrons to produce radio waves which, in turn, interfered with the British army's radar. When the war was over, Hay and other scientists immediately began the study of radio waves emitted by the sun and other celestial objects.

Today we are most aware of heightened sunspot activity when magnetic fields interfere with TV and radio signals and other communications mediated by satellites orbiting Earth. But, if you've ever seen the aurora borealis, or northern lights, you've witnessed another way sunspots affect the planet. Electrons and protons pour from storms on the sun's surface and speed through space. The magnetic field surrounding Earth traps some of these charged particles and holds them in two vast radiation belts that surround the planet like doughnuts anywhere from 250 to 40,000 miles from the surface. The particles in the Van Allen Belts, as they are called, swirl around Earth along the magnetic field lines, chugging back and forth between the North and South Pole. Sometimes, because of changes in the planet's magnetic field, these particles slip down and collide with atoms in the atmosphere. By exciting atmospheric atoms and molecules in the same way that an electric current causes the gas in fluorescent tubes to glow, these cosmic rays paint enormous streamers of radiant color across the sky.

OTHER PARTICLES

Our planet is constantly bathed in protons and other particles that are produced inside giant stars and showered over everything in the universe. Scientists have been interested in them for many years, and it's from experiments with these cosmic rays that the study of nuclear particles got its start.

Physicists discovered that the nuclei of atoms could be shattered by one of these racing particles, which shot out all sorts of other particles never seen before. When particle accelerators were developed, the scientists were able to use these machines to shoot protons or other atomic fragments at nuclei to break them up, revealing hundreds of other strange atomic fragments. (See PARTICLE ACCELERATOR, below.)

During these high speed collisions, energy is transformed into matter for a moment, in accordance with Einstein's famous formula—$E = mc^2$. Locked in the nuclei of matter is a tremendous amount of energy (E) which is equal to the amount of mass (m) in grams multiplied by the square of the speed of light in centimeters per second (c^2). The smallest nucleus is packed with surprising amounts of energy which, under the right amount of force, can be transformed into particles or pieces of matter, if even for a tiny fraction of a second.

As confusing as the growing number of particles can sometimes be, it is helpful to know that they can be divided into two simple categories: bricks and bricklayers. Bricks are fragments that make up the structure of matter, and bricklayers are those that are responsible for moving things around and sticking them together.

Analogous to different kinds of bricks are the two building blocks of matter—*hadrons* and *leptons*.

Protons and neutrons are classified as hadrons, from the Greek word for stout or thick. Interacting powerfully with each other and held together by the strongest force in the universe, the stalwart hadrons are the strong basic structure of matter.

Electrons are classified as leptons, from a Greek word for lightweight. Giving architectural detail, the ethereal leptons interact weakly with each other, easily sharing orbits, mediating partnerships among atoms, and leaping to higher or lower ground.

PARTICLE ACCELERATOR

Using a device called a particle accelerator, physicists have been able to unravel the tightly held secrets of the atom by smashing particles against surfaces or other particles and seeing what falls out. The proton, electron, or some other particle is accelerated to nearly the speed of light, and the energy of its motion and the subsequent collision is enough to break the grip of the powerful force that holds that particle together.

You probably have a particle accelerator at home in your television set. A device in the picture tube fires electrons at a charged surface behind the screen to create the pictures you watch. The accelerators that physicists use to study subatomic particles use much higher energies, of course, and the pictures they see, either etched on film or detected by elaborate sensors, are a view of the most elemental components of nature. Magnetic fields contribute to the accelerated energy of the particle and control its path. As the particle's speed increases, so does its mass, in accordance with Einstein's Special Theory of Relativity. (See A REVIEW OF RELATIVITY, page 17.) Particle accelerators are either long tubes lined with a kind of transmitter that boosts the energy of the particles, or circular accelerators that use strong magnets to deflect the particle into a circular path at higher and higher speeds.

During the late 1920s and early 1930s, physicists in England, Germany, and the United States were all working to create devices that would accelerate subatomic particles to energies sufficient to blast them apart. American physicist Robert Jemison Van de Graaff developed a machine that was called "the atom smasher" by newspapers of the time. It was improved upon by Ernest Orlando Lawrence, a researcher at the University of California, who invented the cyclotron in the early 1930s.

Early cyclotrons passed protons through magnetic fields created by D-shaped magnets. Each time the particles went through the magnetic fields, the energy of their journey was increased and their paths expanded into widening circles. Finally the protons were aimed at a target and smashed into particles.

Linear accelerators blast beams of electrons into metal plates or chambers of liquid hydrogen. At California's Stanford University, the linear accelerator runs electrons and positrons (the electron's antiparticle) down parallel paths until they are forced by magnets to curve into each other. (See ANTIMATTER, page 57.)

A combination of the linear accelerator and the cyclotron eliminates the need for the more expensive giant magnets of the cyclo-

Particle Accelerator

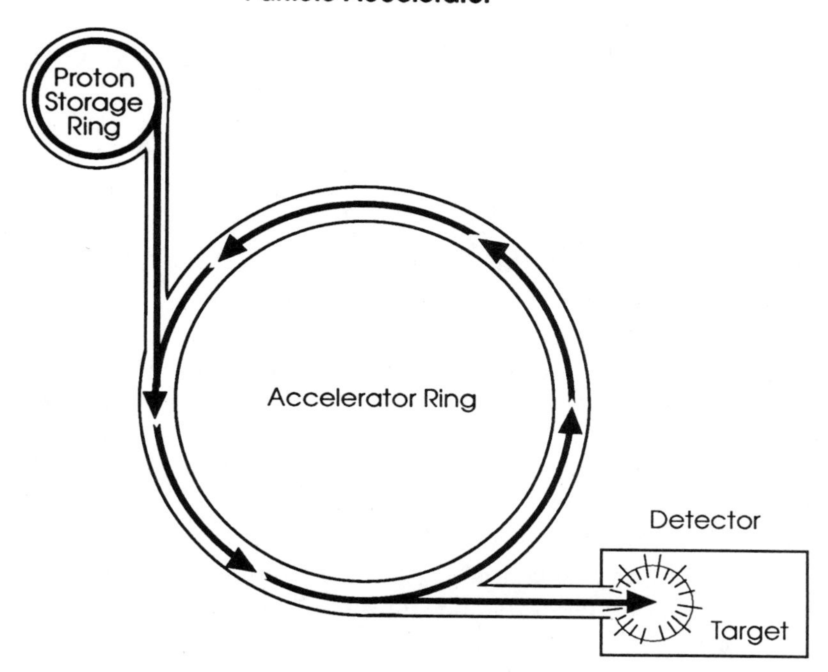

tron, combining less powerful magnets with radio-frequency transmitters to get the job done.

A Superconducting Super Collider, with a 54-mile track, 10,000 superconducting magnets, and ten times the energy of any other accelerator in the world, is now under construction in Waxahachie, Texas. Several other countries, including India, Korea, and possibly Japan, as well as the state of Texas, have pledged their financial support for the project. Nevertheless, the Super Collider will cost billions of dollars, and many people, including a lot of scientists, think such vast sums of money should be spent in other ways. This is not because they doubt the validity of the research, but because they think the money could be spread around over other worthwhile research projects. Other projects that may be more likely to yield economic returns could lose funding, according critics of the Super Collider. Nevertheless, physicists hope to use this accelerator, which is scheduled to go into operation in 1999, to hurl atomic particles at astonishing speeds toward collisions that will pry even more secrets out of the tiny atom. The ultimate goal is find out the most basic information about the structure of matter and how it fits together to make the universe.

UNCERTAINTY PRINCIPLE

To begin to grasp the strange goings on in the hidden world of subatomic particles, you must be like Alice in Wonderland and expect the unexpected. Like the Mad Hatter's tea party, this world is governed by some odd rules.

All of our knowledge relating to subatomic particles is subject to the Heisenberg Uncertainty Principle, named for the scientist, Werner Heisenberg (1901–76), who delivered this bad news to the scientific world in 1927. Injecting a very small dose of uncertainty into the practice of particle watching, the principle states that a particle's position cannot be exactly determined at the same moment that its momentum is measured. Momentum is the measurement of an object's tendency to stay in motion. It is equal to the mass of the object multiplied by its velocity.

If we go back to the comparison between the solar system and the atom, it would be like saying, "We know how fast Mars is going, but we don't really know where it is or what it's like," or "Venus is right over there, but it's impossible to say where it's going or how fast." Even so, the physical laws that govern the solar system are more predictable than the rules of quantum mechanics.

Furthermore, when scientists measure an enormous object like a planet, it doesn't seem to be affected in any way. It goes on ploughing through space as it has for billions of years. Any act of measurement requires an expenditure of energy that will have an effect on the thing measured; that effect is usually so small that it can be ignored. But the very act of observing subatomic particles changes them. That's because subatomic particles are so small that measuring one with even one photon changes its course. Scientists might zero in on the speed of the particle but not be able to comprehend the particle itself. Or, in looking as minutely as possible at the particle, its movement will be somehow compromised. It's comparable to evaluating Mars by bounding another planet into it. You can be sure that Mars would be changed by the process.

QUARKS

The strong materials that are the building blocks of matter—protons and neutrons—are themselves made of sturdy little bricks called quarks so stubborn that physicists are unable to break them out of the building blocks to look at just one. If you think about it, that's the kind of unyielding reliability you want in a good building and the other components of your world.

Using particle accelerators, physicists have broken matter down into what they believe are its most elemental parts. The search for these building blocks of matter had seemed to reveal evidence of disparate particles, but in 1964 American physicists Murray Gell-Mann and George Zweig, each working independently, suggested that many of these particles were just different manifestations of three basic components that differ according to the direction of their spin and the variations in their orbits. Gell-Mann called these particles quarks. Zweig's name for them—aces—never caught on.

The name "quark" is taken from a line in James Joyce's novel *Finnegans Wake*: "Three quarks for Muster Mark!" The three quarks may refer to the three children of the main character in the novel, or three quarts of beer, or even, according to one suggestion, bird droppings. We must assume, however, that if Gell-Mann was thinking of bird droppings, he was not casting aspersions on the origins of the universe. Perhaps the name merely reflects the ubiquity and richness of this most elementary component of matter.

Now scientists think that quarks come in at least six different "flavors," with each one capable of being three different colors. Where quarks are concerned, flavor and color have nothing to do with taste or the wavelengths of light that we perceive as color. The flavors of quarks, that carry the electrical charge of the particle, are up and down, top and bottom, strange and charm. Color refers to the property analagous to the electrical charge of the particles, which glues together different combinations to make up the other particles in the nucleus.

Quarks have fractional electrical charges that, when combined, must add up to a positive, negative, or zero charge, the currencies of atomic activity.

Strangely, the force that holds quarks together increases as they move away from each other. As a result, no naked or isolated quarks have ever been seen; they are locked up in groups in the nuclei of atoms.

It is thought that just after the Big Bang, these components of the new universe were suspended briefly in a kind of seething broth, fondly referred to as quark soup. As the soup pot, or universe, cooled down and particles stopped boiling and bumping violently into each other, the ingredients in the soup began to organize, revealing new forces that acted on the particles. The "strong force" fused quarks into protons and neutrons, the nucleus of the atom, while electromagnetism drew leptons into a wagon train of electrons surrounding the nucleus.

These subatomic campfires ultimately hooked together, by means of their electrons, to create larger structures, the molecules of which our universe is made.

NEUTRINOS

The neutrino was proposed to explain a mystery. When certain radioactive atoms decay, a neutron converts to a proton and an electron. But the transaction, called beta decay, seems to violate an important rule because some of the energy is missing. The First Law of Thermodynamics states that energy cannot be created or destroyed. So, where has it been?

In 1931, physicist Wolfgang Pauli (1900–58) theorized that a particle he called a neutrino was, in essence, the bag man in the beta decay energy heist. It was well disguised and impossible to find because it had no electrical charge and almost no mass, which meant that it would not interact readily with other matter, the usual dead giveaway.

Twenty years after it was first suggested, the neutrino was actually observed and identified by physicists. It wasn't easy, because the wily neutrino slides right through most things, including Earth. Neutrino, which means "baby neutron," is the agent that carries away extra energy from the breakup.

Neutrinos, which were first detected in a nuclear reactor, are now observed and studied at particle accelerators. Scientists believe that they are also a plentiful product of fusion inside stars and, while each one has almost no mass at all, there should be lots of them shed by all the stars in the universe.

So neutrino astronomers have built strange underground detectors in order to catch some of the neutrinos flooding out of the sun and passing through Earth. Tanks of drycleaning fluid or water housed in mines and caves deep in Earth are monitored for the rare interaction between a neutrino and the chlorine or water atoms in the fluid. The tanks are hidden underground to protect the fluid from cosmic rays whose interactions with the chlorine or water would confuse the results. Some neutrinos have fallen into the trap, but not as many as scientists had predicted. They are testing other substances that might serve as better neutrino detectors.

By observing solar neutrinos, scientists can learn more about nuclear reactions in the sun, and perhaps solve another mystery—the case of the missing mass.

GAUGE BOSONS

If protons, neutrons, and electrons are the building blocks of matter, gauge bosons are the muscular, no-nonsense brick layers that carry force. Photons move electromagnetic energy (X rays, light, heat, and so on). *Gluons* act as mortar for the quarks that are the building blocks of protons and neutrons. Part of their job is preventing the positively charged protons from repelling each other. It's an important job, keeping all matter as we know it from disintegrating into a cloud of particles. Other force-carrying particles, called *intermediate vector bosons*, are responsible for radioactive decay.

Scientists think there must be another particle, dubbed the *graviton*, that transmits the force of gravity, although they have no proof. Despite the fact that we feel and see the evidence of gravity every day, it is the weakest of the known forces.

GRAVITATIONAL WAVES

If gravitons are real, then they probably exist, like all matter, as waves as well as particles. Albert Einstein predicted that gravitational waves should affect matter in the same way that electromagnetic waves affect electrons and other charged particles. Scientists are developing instruments that they hope will detect and measure gravitational waves emanating from pairs of neutron stars locked in orbits around each other.

Einstein described space as a medium that is bent and rippled by the objects swimming through it. The distortion of this fabric travels outward from the objects as waves, which weaken as they move along but should nevertheless be detectable by a sensitive measuring device.

So far, all that is known about the universe has been gathered by studying the electromagnetic radiation emanating from the sky. Light, infrared, radio, and other energy waves arose in the universe about a million years after the Big Bang. Scientists believe that gravitational waves have been roaming the universe since the very beginning of time. If they can capture some of this energy and study it, they may learn entirely new things about the formation and development of the heavens.

Unlike light and other electromagnetic radiation, gravity waves roll right through all the stuff (stars, galaxies, gas clouds, dust, and planets) that occupies the universe. A gravity telescope or detector should be able to pick them up. The problem is that gravity waves are weak and will be difficult to detect.

A gravitational wave detector, called the Laser Interferometry Gravitational-Wave Observatory (LIGO), has been proposed and partially funded by the United States government. It would give astronomers yet another tool for studying the contents of the universe. Japan and several other countries are also planning gravitational wave detectors.

Scientists have observed evidence of gravitational waves in the behavior of pulsars, collapsed stars that have strong magnetic fields, and their companions in binary systems. When pulsars are locked in an orbit with another star, they race around in gradually decreasing orbits at high speeds. As a pulsar's orbit sinks toward its companion, it loses energy at a rate that is consistent with Einstein's prediction. Scientists believe this energy is carried away by gravitational waves.

LIGO will look first at neutron stars in binary systems, searching for the gentle perturbations caused by gravitational waves, like a fine net might billow with the slightest waves on a calm sea. Consisting of intersecting laser beams connected to a device that will notice tiny changes in the way the beams intersect, the detector will measure any changes in the intensity of light at the crossroads where the lasers meet. If a gravitational wave comes rolling in, it will jar the mechanism and cause the beams to intersect in a slightly different way.

Someday such detectors may be able to find black holes whirling around each other in a self-destructive dance that sends gravitational waves rippling billions of light years through space and time. And that might not be the strangest thing this new telescope uncovers. Scientists predict many surprises hiding in the folds of the universe.

ANTIMATTER

Incredible, but true. Not only is matter composed of layers of particles, like boxes within boxes, but it is also mirrored by antimatter, which is composed of layers of antiparticles, also arranged like boxes within boxes.

For every existing particle there appears to be a mirror image

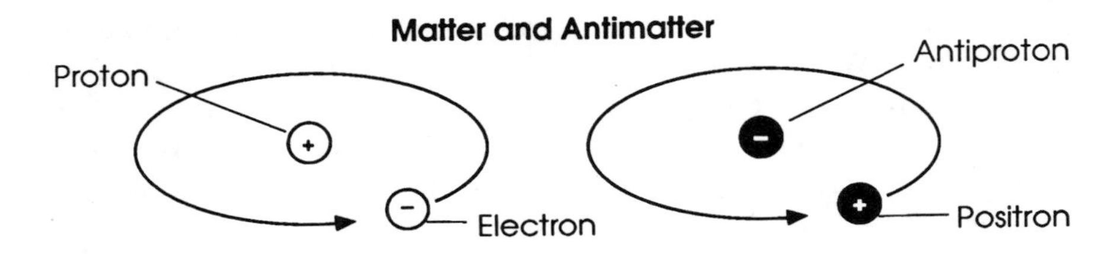

with an opposite charge. For example, for every electron, with its negative charge, there is a *positron*, or positively charged particle. But a tinge of violence colors this tale. When particle meets antiparticle, they are at first attracted, but then they annihilate each other in a burst of gamma rays.

Scientists have observed that in the conversion of matter and antimatter pairs into energy, the mass of the particles is conserved in some way. It doesn't simply disappear. The resulting gamma rays contain the amount of energy that is equal to the mass of the two particles that have been annihilated. The electrical charges cancel each other and disappear.

The existence of antimatter was proposed in 1929 by British physicist Paul Dirac (1902–84). Several years later two American physicists observed odd behavior among the particles they had gathered from the shower of cosmic rays. When particle accelerators came into use, the number of observed antiparticles grew with the sophistication of the technology.

You might be tempted to say, "So what?" But the existence of antimatter is more than an intellectual curiosity. It is the basis for a machine that doctors and researchers use to study the workings of the brain and to detect brain tumors, epilepsy, and other problems. (See POSITRON EMISSION TOMOGRAPHY in Part IV, page 226.)

The tremendous energy released when particles of matter and antimatter annihilate each other may someday be harnessed as fuel for space travel, enabling humans to rocket to distant planets and perhaps other solar systems. Today, astronauts use chemical fuels to blast their way out of Earth's gravity and make limited forays into space. Discouragingly heavy amounts of such fuels are required for lengthy trips; however, only 10 milligrams of antimatter would be needed to produce the propulsion energy of 120 tons of the liquid hydrogen and liquid oxygen that are currently used.

Most scientists consider this a far-fetched scenario. There are some serious practical problems associated with using antimatter as fuel, including finding a container to hold it. When it strikes a surface it disappears! It is also extremely expensive to produce. But some researchers predict that if these and other problems are solved, it may someday be possible to travel to the moon in a few hours and Mars in a few weeks.

STRINGS

Einstein looked without success for a fundamental, underlying force that would unify gravity with electromagnetism, the force that keeps electrons spinning around the nucleus of an atom and is responsible for creating both electricity and magnetism. Since he had already shown the fundamental relationship between mass and energy, he was confident that a unifying relationship between gravity and electromagnetism would also be found. Later, physicists studying quantum mechanics, the measurement and behaviors of small particles of energy at the subatomic level, sought what is known as grand unified theories (GUTs).

GUTs are mathematically supportable theories that posit a common thread among the four forces of the universe: gravity, electromagnetism, the strong nuclear force (which holds together the nucleus of atoms), and the weak force (which influences radioactive decay in some atoms). GUTs would tie all these forces of the universe into one neat package that would help explain the beginnings of the universe and the functioning of the material world today. They are an important and elusive goal of theoretical physicists, who believe that the secret lies somewhere in the opening act of the universe, just after the Big Bang, when the forces were probably still joined.

String theory is one such attempt to unify these forces, which don't always seem to have a common thread. The infinite number of energy particles that physicists have found deep in the universe of the atom may not be separate points or spinning spheres of energy but, rather, parts of one long particle that vibrates like the string of a violin. Photons, quarks, electrons, and antimatter may just be different manifestations of that vibrating string.

Although physicists had previously managed to explain a link among electromagnetism and the weak and strong force, gravity resisted all urgings to join in the fun. However, string theorists have described a manifestation of the string that duplicates the behavior of gravity's elemental particle—the graviton. The four forces may share other such particle behaviors.

But, where's the catch? In order to justify string theory mathematically, it is necessary to believe that there are not just four dimensions to the universe but as many as six more. We don't perceive these dimensions the way we perceive space and time, because they are tightly coiled up and hiding far below the subatomic level.

Not everyone in physics buys this theory. Work continues at physics departments and particle accelerators around the world. Stay tuned for developments.

FUSION

Fusion is the process that lights the fire deep in the bellies of stars. In such a furnace, burning at millions of degrees, hydrogen atoms shake off their electrons and other particles, creating a plasma or soup of atomic fragments. In all the confusion, some of the stripped-down hydrogen nuclei bash into each other and fuse, shooting out more particles and energy. It's this thermonuclear reaction in our own star, the sun, that warms and enriches our planet.

Fusion is also what happens in an uncontrolled way in the explosion of a hydrogen bomb. Having already learned to unleash the destructive power of nuclear fusion, researchers are attempting to domesticate the process into a safe and reliable source of energy. It holds great promise. The fuel source is found in sea water, which covers 70 percent of our planet, enough to keep billions of us in energy nearly forever.

Most of the world's energy is now derived from combustion, a chemical process that breaks the grip a molecule has on its electrons, releasing energy in the form of heat and light. As a power source it's small potatoes, though, compared to the energy that binds together the nucleus of an atom. Although this "strong force" has a very small range, it is a million times stronger and much harder to break than chemical bonds. But the effort may be worth it; the energy payoff is also a million times that of combustible fuel.

The fusion reaction that scientists are trying to perfect is not exactly like that which happens in stars. The sun, which is enormous, plods along slowly, burning its plentiful supply of hydrogen at 15 million degrees Celsius. The collision and fusion of hydrogen nuclei is infrequent enough to ensure that the sun's fuel will last for another 5 billion years.

In order to produce useful amounts of energy with a much smaller fuel supply, scientists on Earth need to use temperatures twenty times that found inside the sun. Even so, the single proton that is the nucleus of a hydrogen atom is frustratingly resistant to fusing with other protons, which are positively charged particles that repel each other once they move out of the binding range of the strong force. (Remember that opposite charges attract, while like charges repel.)

Two cousins of the hydrogen atom seem to be more willing participants in the fusion process. Hydrogen, the H in H_2O, the most prevalent element in the universe and the fuel that fires the sun, has no neutrons in its nucleus. But two isotopes of hydrogen, having extra neutrons in their nuclei, are more willing participants in the fusion process. Deuterium, which has one neutron, is relatively common in seawater, occurring once in every 6,700 hydrogen atoms. Tritium, another isotope of hydrogen, has two neutrons. It does not occur in nature but can be made inside nuclear reactors. Fusion between deuterium and tritium happens more quickly and at a lower temperature.

When two such nuclei smash together, spare parts go flying. The energy produced by fusion is the energy of these departing fragments. This is what Einstein was talking about when he said $E = mc^2$: Energy equals mass in grams times the square of the speed of light. There is a lot of energy locked up in the tiniest scrap of mass, but it requires tremendous heat and pressure to set it free.

These high temperatures create a problem, however. What sort of earthly container could hold the hot soup in which fusion occurs? Temperatures of hundreds of millions of degrees would destroy anything that exists. So scientists have devised a way to keep the plasma suspended in a doughnut-shaped trap created by magnetic fields. Then they blast it with bursts of current or beams of electrically neutral particles to heat it up.

Using such a magnetic fusion device in 1991 at the Joint European Torus experiment in Oxfordshire, England, researchers were able to achieve fusion between deuterium and tritium nuclei, producing 1.7 million watts of power for almost a second. Because fusion reactors produce such high temperatures, they can operate only for brief periods in order to protect the materials in the reactor.

Inertial fusion, an alternative method, can be accomplished by blasting pellets of fuel with lasers. By carefully focusing ten or more laser beams on the surface of a symmetrical pellet isolated in a glass bubble, scientists are able to compress the fuel to a fraction of its original size. Heat is created by this compression, and in such close quarters, excited nuclei bash into each other and fuse.

So far, fusion reactors use more energy than they produce. (Fusion achieved at the JET reactor in England required ten times more power than it produced.) But at least the fuel—water—is cheap. Tiny amounts of material will yield huge amounts of energy. There's enough deuterium in a cup of water to keep a car going for ten years or more.

Does it sound too good to be true? Maybe so. But billions of dollars are being spent on this high-stakes research, and investigators are getting closer every day to the break-even point, when energy created equals energy expended.

Some critics urge caution; the notion of clean and safe energy is compromised by the fact that tritium, an efficient power source, is radioactive, requiring careful handling and the possible decontamination of the reactor building. Still, useful fusion power may be possible, experts say, in about fifty years. None too soon, when you consider that the world's petroleum reserves will be running seriously low in a hundred years.

COLD FUSION

Considering the problems associated with hot fusion—among them the need for staggering amounts of energy to create temperatures higher than those inside the sun, and a complex containment system such high temperatures require—it is not surprising that hot fusion consumes more energy than it produces. So the possibility of achieving fusion in a room-temperature jar of water, using relatively simple and inexpensive laboratory equipment, is pretty enticing.

When B. Stanley Pons, a professor of chemistry at the University of Utah, and his British colleague, Martin Fleishmann from the University of Southhampton, announced in March 1989 that they had done just that, people really got excited.

The theory behind the process used by Pons and Fleishmann involves putting a negative and positive electrode into a container of *heavy water*. Whereas the hydrogen in water has one proton in its nucleus, the H in heavy water is replaced by its deuterium isotope, which has a neutron in addition to the single proton. A current is passed between the electrodes, breaking apart the heavy-water molecule and releasing positively charged deuterium nuclei. These are absorbed by the negative electrode, which is made of palladium, a soft, silvery metal that is used in electronics and dental fillings. The crystalline structure of palladium allows the deuterium nuclei to crowd closely together in the empty spaces in its structure. By forcing the deuterium nuclei so closely together, the process would overcome the strong force and allow fusion. The production of heat and the presence in the water of helium and other products of hydrogen fusion led Pons and Fleishmann to believe that their experiment was a success.

Critics suggest that Pons and Fleishmann set off on the wrong foot by not adhering to the normal standards for sharing scientific information. They announced their results at a press conference, instead of publishing in a scientific journal and sharing the details of their experiment with colleagues.

Scientists raced to duplicate the experiment, however, and most were not able to support the Utah team's findings. Others did report the production of heat and the presence of fusion products, but critics have suggested that the presence of helium, tritium, and other products of fusion was due to contamination from some other source, not the result of real fusion.

Millions of dollars flowed into cold-fusion research, fueled by the intense interest of governments, industry, and a general public weary of the environmental and political problems associated with petroleum energy. But when experiment results were inconclusive, the furor and the funding began to die down, despite urgings from some scientists who believe that cold fusion may still work. It may not be fusion, as it's currently defined, advocates say, but some other atomic reaction waiting to be discovered. The experiments continue.

FISSION

Fission is the opposite of fusion. When the nucleus of a heavy element like uranium or plutonium is bombarded with a neutron by means of a particle accelerator, the nucleus breaks apart into two smaller particles, releasing extra neutrons and radiant energy, or heat. The extra neutrons cause other nuclei to split apart setting up a chain reaction that can end in a massive explosion if not somehow controlled.

Fission was first recognized in a Berlin laboratory by German physicist Otto Hahn (1879–1968) and Austrian physicist Lise Meitner (1878–1968). They used a kind of particle accelerator to bombard uranium with neutrons and investigate the results. Meitner, who was Jewish, was forced to escape to Sweden to avoid Nazi persecution, and German chemist Fritz Strassmann became Hahn's partner. Hahn and Strassmann struggled to identify the much lighter elements that seemed to emerge from the bombardment. Hahn described this in his letters to Meitner and it was she who suggested that the uranium nucleus had actually split in two. She named the process fission.

When the news of these results reached other physicists, they quickly went to work trying to duplicate the experiment. The implications of the discovery—that the energy and extra neutrons released by fission could be used in a powerful bomb—quickly occurred to several scientists. Fearing that Nazi Germany would develop such a weapon first, Hungarian-American physicist Leo Szilard (1898–1964) convinced Albert Einstein (1879–1955), who had left Europe for America, to write to President Franklin Roosevelt warning him of this possibility. In a letter written on August 2, 1939, Einstein urged the president to authorize research so that the United States could develop the nuclear-fission bomb first. On December 6, 1941, Roosevelt set into motion a project that would soon bring the world into the nuclear age.

In 1942 scientists at Iowa State College at Ames began working to purify enough uranium to fuel a chain reaction. Such a facility was built at the University of Chicago under the leadership of Italian physicist Enrico Fermi (1901–54). The uranium used in the reactor was layered in a stack with graphite serving as a moderator, a substance that slowed down the neutrons escaping from the fission reaction. Slowed–down neutrons do a better job of splitting apart the uranium nucleus. Holes cut into this stack of material were fitted with rods made of cadmium, a soft, metallic substance that absorbs neutrons. In order to set off the chain reaction, the physicists sent a stream of neutrons into the stack to begin the fission process and then gradually withdrew the rods. On December 2, 1942, the first self-sustaining fission reaction occurred. The first bomb was successfully tested at Alamagordo, New Mexico, on July 26, 1945. Two others were subsequently dropped on Hiroshima, Japan, on August 6, 1945, and Nagasaki a few days later, bringing World War II to an end.

NUCLEAR POWER

In the early 1930s scientists discovered the powerful potential in the nucleus of the atom. Experiments quickly released that power in the form of nuclear weapons during World War II. After the war, the bright promise of nuclear power in peacetime gave rise to all sorts of ideas for using this new resource. But ultimately its cost and the dangers of containing its radioactivity caused most of these hopes to fizzle. A notable exception is the nuclear-powered submarine. While nuclear power is also used for research and medical diagnosis, its chief use today is in the production of electricity.

At first glance it's easy to see why nuclear power plants appeared to be the answer to the world's energy problems. The energy released in the fission of uranium atoms is wildly out of proportion to their mass. One ounce of uranium can provide the same amount of energy as almost 100 tons of coal.

Uranium is a relatively common but thinly scattered component of the Earth's crust. In order to use it to generate power it is necessary to crush tons of uranium ore and use chemicals to dissolve out the small amounts of pure uranium in it (often 1% or less). Only a tiny fraction of it is Uranium-235 (U-235), the type needed for fission reactions.

The nuclei of U-235 atoms are easily split when bombarded by other atomic particles. When struck by a stream of neutrons, the nuclei fracture into new elements that weigh less than the original nucleus. The missing mass is converted into energy. It is this energy that is harnessed to heat water and create steam and generate electricity in a typical nuclear power plant. Two extra neutrons are also released in the fission of each U-235 nucleus; they strike other U-235 nuclei, setting off a chain reaction.

The uranium fuel itself is shaped into pellets which are packed into metal tubes and placed inside the reactor core along with the water. Rods made from a neutron-absorbing substance are included with the fuel rods. By slowing down the neutrons and also taking some of them out of the game, so to speak, this material—called a moderator—improves the rate and controls the number of fission reactions that take place.

In most nuclear power plants in the United States, ordinary water acts as the moderator and also is used to cool the reactor core and transfer its heat to the turbines to generate electricity. In a boiling water reactor the kinetic energy of the fragments shooting from the nuclei heat the water in the reactor to boiling and the steam is piped away to a turbine. In pressurized systems the water in the core is pressurized to keep it from boiling. It is piped within a closed loop through a separate vessel of water which boils and produces the steam that drives the turbine. Water is not necessarily the most efficient moderator and coolant, however, because the hydrogen in water molecules (H_2O) captures neutrons and forms a material called deuterium. This process results in a very low energy yield. But water's advantage, is that it is plentiful and can serve a dual purpose—moderating and cooling.

An alternative is to use heavy water, with deuterium atoms in it. Heavy water absorbs fewer neutrons and reactors that use heavy water as a moderator and coolant can use natural uramium, instead of enriched uranium, for fuel.

Gas-cooled reactors use graphite, a form of pure carbon, as the moderator and a gas, usually helium or carbon dioxide, as the coolant. Helium's advantage over water is that is can be heated to much higher temperatures, and as a result, reactors cooled with it offer a much higher thermal efficiency than water-cooled reactors.

The reactor core, along with the steam generator and a pump that circulates a coolant around the fuel, are all contained in a steel vessel to prevent the escape of radioactivity. This containment vessel is also surrounded by concrete, which also absorbs radioactivity.

It's reasonable to think that the limited amount of U-235 in the world would soon put an end to nuclear power plants. However, of the 200 or so products of the fission reaction, some are also useful as reactor fuels. Certain reactors—called breeder reactors, for obvious reasons—end up producing more fuel than they consume.

For example, one kind of breeder reactor is stocked with fuel containing plutonium and U-238, a more abundant form of uranium that is not, by itself, fissionable. The fuel is more densely packed together and the moderator is left out. Neutrons leaking from this core strike a lining made of U-238, converting some of it into a fissionable plutonium. These fast breeder reactors, as they are called, are usually cooled by liquid metals because of the high temperatures generated by the fission reaction. Thermal reactors, if loaded with the appropriate fuels, can also be used as breeder reactors. Until recently, the United States military used breeder reactors to produce plutonium for nuclear weapons.

Plutonium is a dangerously toxic radioactive substance, and the production and handling of it and other fission products is a serious problem for society. Today only about 20 percent of the electricity in this country is produced by nuclear power plants. An accident in 1979 at the Three Mile Island nuclear power plant in Pennsylvania, even though there was not a heavy release of radioactivity, only confirmed what Americans already feared—the risks associated with this kind of power generation may be too great. Indeed, the number of new nuclear power plants being built throughout the world decreased from 200 in 1980 to 50 in 1991.

The accident at Three Mile Island occurred when a coolant pump malfunctioned. Without the coolant, the fuel rods began to melt. Radiation was prevented from escaping in catastrophic quantities by the concrete that surrounded the containment vessel. In a complete and unchecked meltdown the fuel rods could eat through the containment wall and continue into the Earth. This is known as the China Syndrome, because of the popular notion that the molten fuel would eventually dig through to China. While that would not really happen, the damage the runaway fuel would do to the environment would be devastating.

A more serious accident occurred in 1986 at Chernobyl, in the former Soviet Union, when an explosion and fire caused a meltdown. In this case, the reactor core was not contained in concrete. The explosion literally blew the 1,000-ton steel lid off the reactor, spewing radiation into the air. Unlike reactors in the United States, in which the fission reaction slows when coolant is lost, the Soviet reactors are designed in such a way that a leak could lead to increased power and runaway nuclear reactions. In addition to what scientists determined were serious design problems, there were operator errors that led to the accident that killed 30 workers at the outset and contaminated large sectors of the Ukraine, one of the richest agricultural areas in that part of the world.

Besides the very real dangers of operator error or safety system failures—both of which happened at Three Mile Island and Chernobyl—there is the problem of what to do with the radioactive waste that nuclear power plants produce. The fuel, uranium atoms, is not as radioactive as the smaller atoms that are produced by fission. Plutonium, for example, has a half-life of almost 25,000 years. (A half-life is the time it takes half a given amount of an unstable element to decay or throw off sub-atomic particles until it is stable.) Some radioactive elements, like Iodine 129, are dangerous for millions of years.

The spent fuel is usually stored in cool-water tanks near reactors. Some countries bury their radioactive waste in what they have determined are geologically stable areas. The United States government plans to create a long-term storage facility in an underground repository. Yucca Mountain, one hundred miles northwest of Las Vegas, is the leading candidate for this facility. Scientists are currently evaluating the geological formations in this desert area in order to determine its suitability for long-term nuclear waste storage. A final decision may not be possible until well into the next century. There are many challenges facing such a project and the likelihood of its success is extremely controversial. The most important is forecasting the long-term geological stability of a site. It may be a dry quiet hole in the ground today, but what will it be like in a thousand years or more? Climate changes, earthquakes or volcano eruptions could cause serious problems for future generations if scientists do not find a method of decontaminating the waste.

CHAOS

You see it in the plume of smoke that rises straight and uncomplicated from the campfire and then breaks suddenly into curls and eddies before it blends into the sky. It is in the languid river that splinters into turbulance as it tumbles over rocks. It may even account for wild fluctuations in the stock market. But unlike our everyday understanding of the word, chaos, in the technical sense, is not total disorder. It is orderly behavior disguised as anarchy.

Scientists have always known that there are mysteriously chaotic elements in life, like the turbulence in water and weather, although they never thought these patches of apparent disorder were a natural part of things. Instead, they looked for rhythms and patterns in nature, assuming that this measurable, periodic behavior was the governing force. This is not surprising. Our understanding of the universe has been shaped by Newton's orderly laws of motion and theory of gravity.

But could it be that there is another game plan, both unpredictable and unstable, that is managing the business of the stars and the beating of our hearts? Are even the simplest systems haunted by a ghostly chaos that we cannot easily see? Maybe so.

A good example is the heartbeat, which you expect to be regular and predictable, right? But researchers have discovered that the normal heartbeat is not regular at all. Hidden within its superficial order are complex fluctuations that resemble the chaos being discovered in other systems.

The rules of behavior in a chaotic system include "extreme sensitivity to initial conditions." That means the system behaves predictably at first, a creature of cause and effect; then the smallest event can send it tumbling off on a complex path far from home with no way back. It rolls with the punches.

It makes sense that in human physiology, this adaptability is one hallmark of a healthy system: People who are on the brink of a heart attack seem to have lost this chaotic flexibility. Research into this phenomenon may change the way patients are monitored for the effects of heart disease, drugs, and aging, giving physicians a head start on solving problems before it's too late.

The mathematics behind the theory of chaos have been known since the turn of the century. Henri Poincaré (1854–1912), a French mathematician and physicist, was stunned to uncover the possibility of chaotic behavior in the solar system, which Newton's physics had declared perfectly stable. Poincare discovered that Newton's equations could not predict stability in a system of more than two bodies. He was so upset that he turned his back on the idea while other researchers ignored his findings altogether.

While Poincare and other early chaos detectives knew something weird lurked behind the orderly face of Newton's universe, they couldn't do much about it. They didn't have our high-speed computers to do the seemingly endless calculations necessary to reveal chaotic patterns. Like early mapmakers who drew dragons beyond the edges of the known world, scientific explorers thought their glimpses into chaos revealed monsters. Eventually, though, researchers with interests in both mathematics and other sciences began to explore the chaotic landscapes hidden in repetitive mathematical problems, as well as in natural systems.

In 1960, Edward Lorenz, a meteorologist at the Massachusetts Institute of Technology, created a simple weather system on his computer and not only discovered a pattern to the chaos of weather but also graphically illustrated the sensitivity of the system to even the tiniest effect. Lorenz got disproportionately different results in his calculations after rounding off a number by a tiny fraction. He came to a conclusion, now famous, that weather is so unpredictable because something as small as the beat of a butterfly's wing in Brazil can cause a tornado in Kansas in as little as two weeks. This conclusion dashed any hopes meteorologists had of making long-term predictions of the weather. It also drove home the notion that even very simple systems with as few as three "degrees of freedom" (or independent variables) are unpredictable.

Computers have enabled researchers like Lorenz to plot complicated multidimensional diagrams of systems, a kind of mathematical topology that Poincaré had begun years before. Patterns emerged, and then hauntingly beautiful computer graphics were created—almost like maps that scientists could study to learn more about the behavior of chaotic systems.

Another important development eventually contributed to the understanding of chaos. Benoit Mandelbrot, a researcher at Yale University, had rediscovered the fractal, an old idea that resurfaced in these tracings of complex systems. Fractals are fractional images of something, in which a part looks exactly like the whole thing. The classic example is the pattern of England's coastline, which is remarkably similar from any perspective, from satellite photo down to the ant's-eye view. A fractal is impossible to measure definitely because the self-similar image, full of angles and bumps, goes on and on. Researchers find fractals—the footprint of chaotic systems—in a remarkable number of natural things, including frost crystals, rivers, the human nervous system, and clouds.

And finally, as with any intriguing new idea, evidence of chaos has been spotted in all sorts of unlikely places. Chaos has become a useful metaphor in art, politics, economics and religion. It is also making strange lab partners of previously separate sciences like physics and biology, creating new disciplines in which researchers paint more detailed pictures of life's delicate rhythms. The bad news is that they may never be able to do anything about predicting the weather.

II

THE CITY OF THE CELL

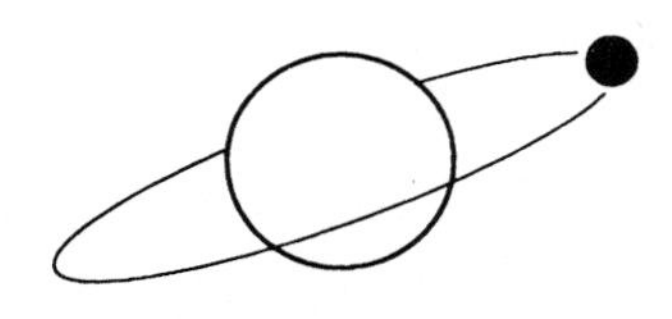

DESPITE OUR FEELINGS of superiority, we humans are cut from the same plain cloth as every other living thing. Our DNA, like that of field mice, tomatoes, bacteria, and the neighbor's yappy dog, is determined by the same four molecules, called bases, which code for the same amino acids that constitute the proteins that form blood, bacteria, and fur. DNA's chemical building blocks are all arranged in the same unit of one sugar, one phosphate, and one base and are called nucleotides. Three nucleotides, in turn, make a codon. There are 64 different possible combinations of codons, and the way in which they engineer life is essentially the same from critter to critter.

The main difference lies in the way the codons are strung together into a code. The code for human is different from the code for vegetable or whale, for example. But we are made from the same stuff, and some of our DNA is even interchangeable. In light of that fact, it is easy to think that we all came from some common organism that crept from the ancient soup of early Earth and over millions of years achieved the diversity that makes our planet rich. It's not so, but we are still cousins under the cell membrane nevertheless. Our past and our futures are linked together by ancient, invisible bonds of DNA.

The delicate interdependence of all living things may have as much to do with this common past as it does with our current dependence on each other. We each occupy one strand in a giant web of life. Disturb one thread and the entire structure may feel the effect. Some of the concern about human tinkering with the blueprints of life has to do with this interdependence. Critics have worried that scientists will make irreversible changes in the web. They suggest that researchers may be tempted to alter the reproductive cells of humans and make changes that will pass from generation to generation. And they worry that altered forms of life might escape from labs and holding tanks to upset the delicate balance in ecosystems and create unforeseen problems.

Scientists themselves say it's important to realize that genetic recombination happens in nature all the time and is the basis of life on Earth. It's not something scientists invented, but rather, is founded on basic biology.

This is a complicated matter. Scientists tell us that ecosystems are remarkably robust. A tweak in one area of the web can cause the entire system to shift and compensate for the change. As a result, it is sometimes difficult for researchers to predict the consequences of change or pinpoint the cause when an ecosystem actually collapses.

In any case, how is the web of life altered when we remove an organism through forced extinction? How do we change the interdependence of life through desertification, deforestation, pollution, and ozone destruction? We have toyed with the diversity of life through the selective breeding of animals and plants. In fact, nearly every human activity seems to exact a price from the system of which we are a part.

But, as you will see, it is difficult to deny the value of the astonishing research that creates lifesaving drugs and other products that make life better for all of us, sometimes even for animals and plants.

ORIGINS OF LIFE

Living cells emerged on Earth around 3.9 billion years ago and quickly filled the oceans until there were enough to leave fossil evidence. There are tracings of microscopic clumps of cells that date back to 3.5 billion years ago.

Somehow, in the dim beginnings of life on planet Earth, carbon atoms formed bonds to create the giant molecules that are the basic structures of living things. Scientists aren't sure how that happened, but they have some interesting ideas about the process.

Life probably began as a series of chemical reactions in the ocean, which served as a primordial soup pot full of the necessary ingredients. Carbon dioxide, ammonia, hydrogen, and methane had been thrown into the air by the first volcanic eruptions, building a primitive atmosphere. Its gases dissolved into the surface of the ocean and were stirred into a carbon-rich soup, which scientists had tried for years to duplicate in the lab.

In 1953, Stanley Lloyd Miller at the University of Chicago succeeded. In two bottles of water connected by glass pipes, he cooked up a mixture similar to that which might have formed in the oceans. To simulate the lightning that may have jump-started the chemical reactions, Miller passed 60,000-volt electrical sparks through the mixture. After a few days, it turned cloudy red with amino acids, the building blocks of protein and life. Miller also found short pieces of RNA, the chemical code that builds enzymes. Subsequent experiments supported these results.

Researchers also found that ultraviolet radiation, a component of sunlight, in addition to heat like that in the early volcanic eruptions on Earth, could trigger a similar reaction. In this way, nonliving chemicals could be stimulated to form the molecules of which life is made.

Carbon-based, or organic, molecules have important qualities that made them ideal building blocks of life in the primitive oceans of this planet. Carbon atoms bond easily among themselves and with other atoms to make molecules that have the startling ability to replicate themselves. It's thought that the earliest molecules served as templates for the formation of more molecules, setting off a chain reaction that quickly filled the oceans. Their numbers grew exponentially, each molecule gathering up other carbon, oxygen, nitrogen, and hydrogen atoms into new structures that fell away to serve as templates themselves. The ease with which carbon formed associations with other elements gave birth, researchers say, to the molecular complexity and diversity of living things.

Some scientists have suggested that meteorites that contained the same ingredients, including scraps of water, could have planted the seeds of life on Earth. However, the planet may have already had what it needed to get started—abundant water, a hospitable temperature, as well as the necessary elements that had been forged long before in the explosions of giant stars and been gathered into the objects that make up our solar system. (See THE LIFE AND DEATH OF STARS in Part I, page 26.)

Although researchers have succeeded in triggering the formation of amino acids, they have yet to make living cells from scratch.

Experiments have shown that large organic molecules tend to curl up into cell-size spheres that even seem to split apart in a way that resembles reproduction. But they are not alive. All of the mechanisms by which organic molecules organize themselves to perform the functions of real living cells have not yet been uncovered. The small and seemingly simple individual cell is actually a complex organization of millions of molecules with many well-guarded secrets.

The first cells may have formed in the bottom of the ocean, safe from the damaging ultraviolet rays of the sun and from the many chunks of rock that bombarded the planet during those early years. Hydrothermal vents, islands of sulfur-based life systems that exist on the floors of deep oceans and lakes, may have been the site of early cell growth.

Life might also have sprung from between layered sheets of clay, an ancient doughy substance that is deceptively complex. The crystalline structure of clay is capable of storing and releasing energy. Some scientists even think clay is capable of a primitive kind of evolution of mutations in its crystalline structure. It would have made an ideal substrate on which the molecules of life could form.

Noting that cell walls are made primarily of lipids or fats, some researchers have suggested that primitive lipids might have wrapped themselves into bubbles, capturing some of the organic molecules adrift in the primeval sea and serving as protective cells in which life could evolve.

However it happened, scientists are convinced that evolution occurred first as a chemical reaction, laying the groundwork for the biological evolution that followed.

THE CELL

It has become clear to researchers that a key to understanding how life began and survives is learning the secrets of the cell. Every plant or animal is essentially a large community of cells that transports essentials and provides protection, cleanup, communications,

Parts of a Cell

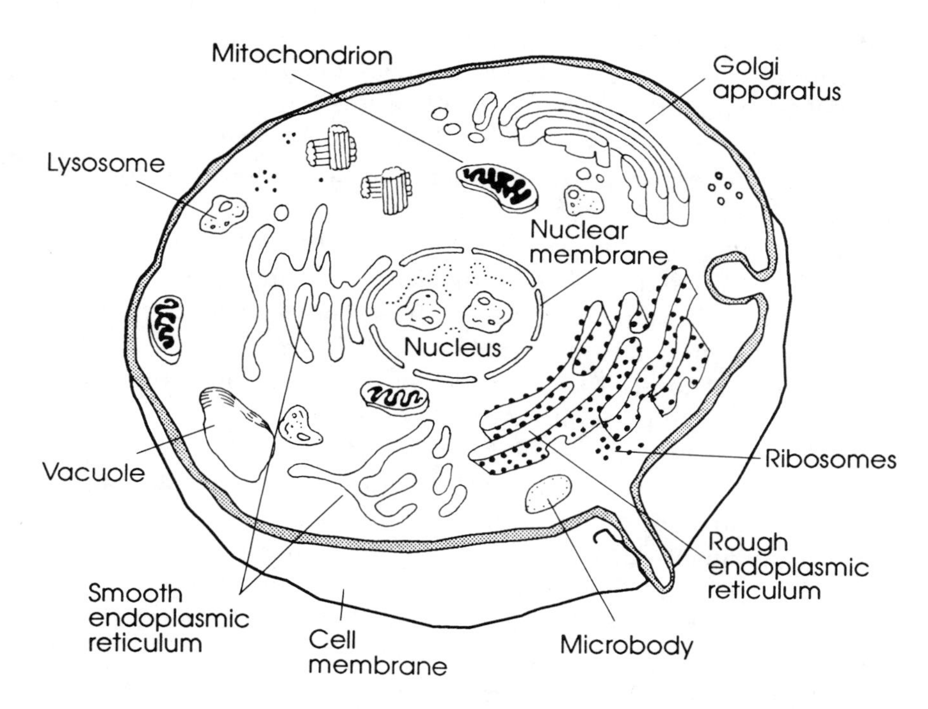

growth, maintenance, and so on. Using carbon, water, and minerals as raw material, the cell produces enzymes, hormones, and other proteins that will do the work. The cell itself is also comparable to a small city, with functions assigned to tiny workers within its walls.

The double-layered membrane of the cell is made of fat and protein molecules. Embedded in this cell membrane are special molecules or receptors that communicate the needs of the cell to the sea of fluids in which it lives. They act like secret passages, each responsible for bringing in a specific molecule. Some substances are able to pass in and out through the cell walls by means of diffusion or osmosis.

Some cells are capable of wrapping themselves around material and sucking it inside, creating a vesicle or capsule that is squeezed off and shuttled away on a built-in transportation system to the location in the cell where it's needed. Some cell membranes—such as those found in mucus membranes and the lining of the intestine—are also equipped with tiny hairlike projections called microvilli, from Latin words meaning "tiny hairs." They increase the surface area of the cell and thus its ability to absorb things.

Inside the cell membrane, the city of the cell is populated by a constellation of little workers, called organelles, that perform the services that keep the cell alive. They can be compared to the workers, factories, and transportation and communication systems that are vital to a working community.

For example, the power plants for the cell are the *mitochondria*; there can be thousands of them in each cell, burning the carbohydrate fuel that is taken into the cell. Mitochondria convert the carbohydrate fuels from the food we eat into ATP (adenosine triphosphate), a molecule that the cell uses for energy. In other words, mitochondria are rather like powerplants burning coal to make the electricity that keeps the city running. If we could isolate and extract all the ATP produced by an adult human body in one day, we could power 1,500 100-watt light bulbs for one minute.

Because mitochondria have double-layered membranes and their own DNA, they are thought to have once existed as separate cells. Apparently, at some time in the dim evolutionary past, mitochondria and our ancestral cells found it mutually beneficial to join forces. This symbiotic relationship probably began when a bacterium was engulfed by a vesicle and pulled into a cell.

Other workers, *lysosomes*, are comparable to fuel refineries and waste-processing plants. They break down chemical raw materials into a more usable form for the mitochondria, and they also digest unwanted substances. A related duty is rather sad. If the city cell comes under siege, as in an accident or illness that deprives it of oxygen, lysosomes release digestive enzymes that kill the cell and its citizens.

Throughout the cell are pleated networks of membrane called the *endoplasmic reticulum*, which serve as a transit system for materials that are being shipped to and fro. A portion of the endoplasmic reticulum is studded with *ribosomes*, organelles which serve as workbenches where proteins are produced.

The *Golgi apparatus*, a specialized part of the endoplasmic reticulum, is a strange-looking organelle that scientists suspected but didn't really see until the advent of the electron microscope. It looks like a stack of plates or bowls and serves as a finishing and packaging warehouse for proteins, the city cell's chief exports. Some of these products are packaged in vesicles, which are transported to the membrane and dumped outside.

Threading through the cells are strings of protein that function like the posts and beams of a building or bones in a body. The *cytoskeleton*, as it's called, gives cells shape, but it also allows them to move and change shapes when necessary. *Microtubules*, or tiny tubes, are also scattered throughout cells, seemingly appearing and disappearing where needed. They act as roadways over which vesicles of chemicals are shipped.

The seat of government and memory bank for this city-cell is the nucleus. Some primitive cells, such as bacteria and blue-green algae, do not have nuclei; they are called *prokaryotes*, which means prenuclear, or having a primitive nucleus that is not bound by a separate wall. Cells that contain a nucleus are called *eukaryotes*, meaning they have a true nucleus. Inside the special wall that separates the nucleus from the rest of the cell are twisted, threadlike bodies called *chromosomes*. Coiled around the protein core of the chromosomes are strings of genes, or instruction codes, that are made out of DNA, deoxyribonucleic acid. DNA is the instructional manual for the operation of the cell and ultimately the larger community.

DNA and RNA

DNA's structure—first described almost forty years ago by scientists James Watson and Francis Crick—is a double helix, a twisted ladder of chemicals. The sides of the ladder are made of alternating molecules of phosphates and a kind of sugar called deoxyribose. The rungs of the ladder are made of the pairings of four nitrogen-based substances called bases: adenine, thymine, cytosine, and guanine (A, T, C, and G). T bonds only with A, C only with G. The basic unit of DNA—one phosphate, one sugar, and one base—is called a nucleotide. Three nucleotides together make the unit of genetic coding called a codon.

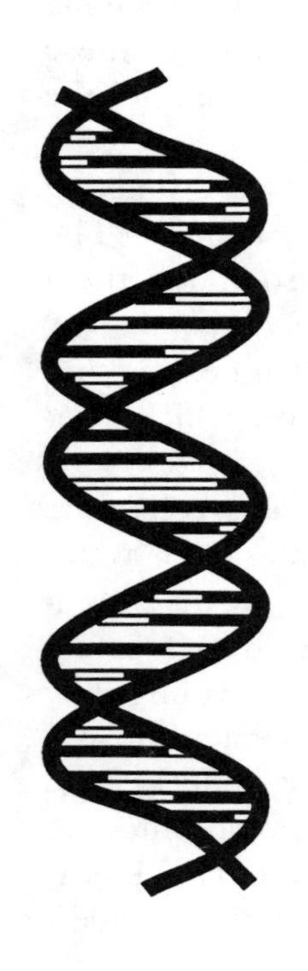

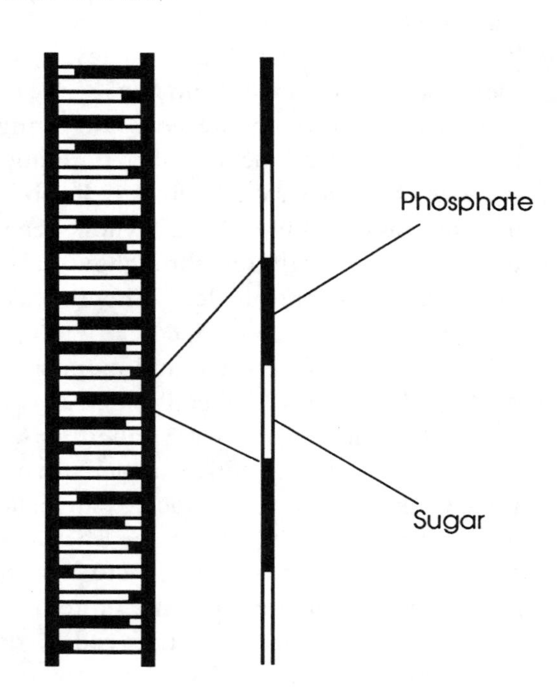

The various ways in which the codons combine create sequences of code, called genes—the recipes for proteins, the building blocks of life. This invisible coded message is coiled up in chromosomes, twisted pairs of threadlike bodies within the nucleus. A chromosome consists of a protein core around which the double helix is wrapped. It gets its name from the Greek words for "colored body," because it readily accepted dye in the laboratory when researchers were trying to discover its secrets.

The simplest living creatures, bacteria, have one single chromosome each. Fruit flies have 8, garden peas have 14, and some butterflies have hundreds. These sets of chromosomes turn tiny blobs of cells into the many wondrous living things that inhabit the earth. Human cells contain 46 chromosomes, which determine eye and hair color, height, genetic diseases, and many other characteristics that make us what we are. Two of the chromosomes, called X and

Y, determine sexes and carry other genetic characteristics that are not related to sex determination.

The human body contains trillions of cells. Each one houses complete instructions for the entire body, although every cell does not use all the instructions it contains in its nucleus, only those important to that cell's job. The coded instructions include some 3 billion different pairings of the four bases A, T, C, and G. The DNA from one human cell, untangled from its spiral staircase, would stretch out to six feet. That's about 17 billion miles of DNA in one human body.

It was difficult for scientists to understand how this incredible molecule managed to do all that it did. DNA sat serenely in the government center of the cell like a wise and aloof queen, never leaving, changing only with infrequent mutations. How did it actu-

Replication by RNA

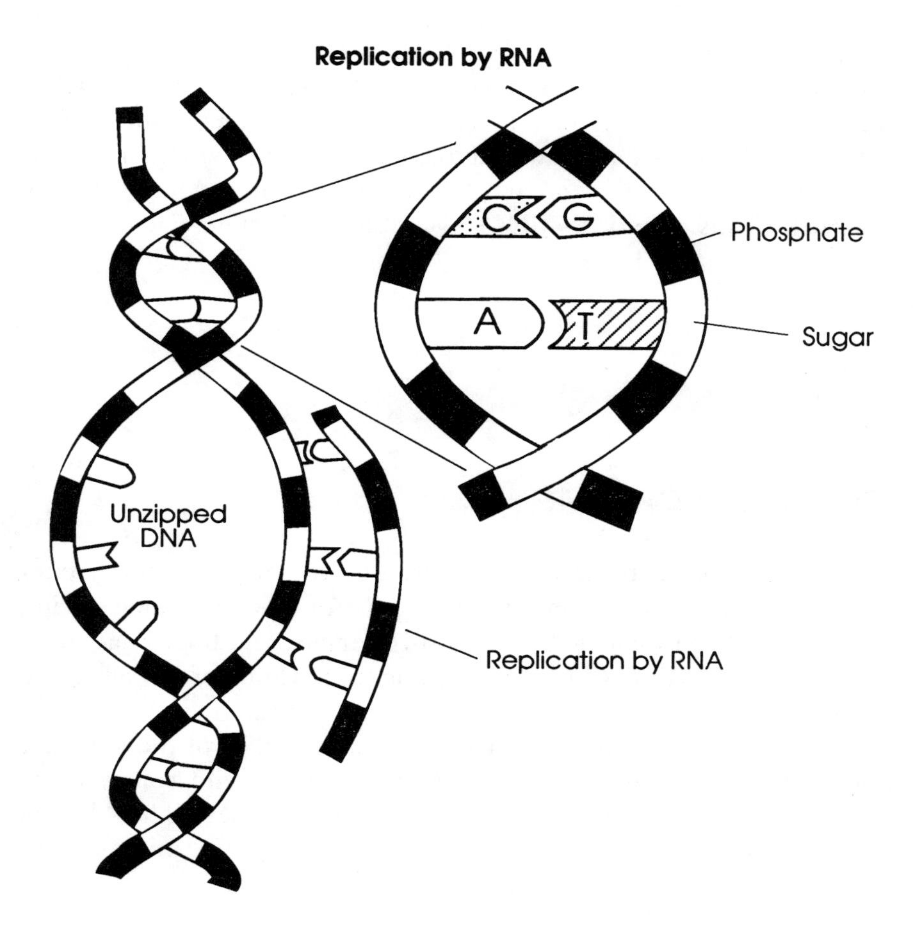

ally create life from its isolation? Clearly, other helpers had to be involved in the task, and they were finally identified.

The others include a kind of scribe and messenger molecule, messenger ribonucleic acid (mRNA), which is, except for a few atoms' difference, a mirror image of DNA. Instead of T, it has the base uracil (U), which pairs with A. Instead of the sugar deoxyribose, the several kinds of RNA have a sugar called ribose. mRNA is the molecule that copies instructions from the DNA molecule and carries them out into the cytoplasm of the cell. In order for this to happen, the DNA must unzip the part of the double helix that has to be copied, in a process called transcription. After the DNA ladder has split apart, the scribe or messenger RNA (mRNA) is assembled from material in the nucleus in a mirror image of the DNA template, and then carries the information out of the nucleus.

Out in the cytoplasm another kind of RNA, called transfer RNA (tRNA), reads the instructions and forages for some of the twenty amino acids that are the raw materials for protein production. Then, on a ribosome—a kind of living workbench that generates energy from the transformation—tRNA assembles the amino acids that are needed to sustain life.

THE HUMAN GENOME

DNA and RNA activity is coordinated by the genome, the master plan governing the construction and function of each living creature. Despite all the obvious differences, the human genome is remarkably similar to the genetic blueprints of other living things.

Scientists have embarked on an exhaustive survey of the genome, not only to find out more about how life began and continues to work, but also to get at the root of many diseases. The Human Genome Project, supported by $3 billion from the United States government, has been called the "Holy Grail of biology," and so far, crusaders from around the world have deciphered 35 million base pairs of A, T, C, and G, specifically identified 5,000 genes, and accurately located 2,000 of them on chromosomes.

Of the four thousand known hereditary diseases, one hundred are now understood well enough to make diagnostic tests possible. Researchers are also trying to identify unique stretches of DNA located every 5,000 to 10,000 base pairs and use them as markers. This will simplify the process of identifying individual genes by placing known landmarks in the field.

A gene is a sequence of code that is transcribed into RNA code which in turn is transcribed into a particular protein, but the human genome is more than just a long list of ingredients; it also includes detailed instructions for the use of the recipes. In a gene, the order of the codons—units of three nucleotides, each of which includes a sugar, a base, and a phosphate—indicates which amino acids should be combined to make proteins. The gene also has sequences of code called promoters, and regulatory regions that restrict where and when the cell produces a particular protein. In this way, the specialized cells that make up the different parts of the body use only the recipes that they need to fulfill their particular function, even though most of them include the whole cookbook for the human body. Researchers expect to find other control codes hiding in the seemingly endless sequences of nucleotides.

Some control codes mark the beginning and ending of important sequences. The instructions for a particular protein may be broken up into subunits, or *exons*, by special interruption codes, or *introns*. In some genes these subunits are combined in different ways to create specific proteins, rather like variations of a recipe. Some sequences include dozens of such interruptions. Some larger interruptions have smaller, separate genes hiding within them, sometimes on the opposite strand of the double helix and reading backwards, although the proteins for which these mysterious little genes provide the code are not known.

There are also plenty of "typos" in the human genome, mutations that have occurred over the lifetime of the human race as well as each individual. Many of these mistakes have no effect, especially since the genome includes double copies of most of the instructions, and the cell has mechanisms for repairing many mutations that do occur. Some mistakes have mixed results. For example, the mutation that causes sickle-cell anemia also provides resistance to malaria. In sickle-cell anemia, the red blood cell is malformed—like a sickle instead of a disc—and contains an abnormal form of hemoglobin, the pigment in the blood that carries oxygen. When the gene is inherited from both parents, it can bring the onset of the fatal disease.

Some instruction codes seem to work correctly only when they are donated by either the mother or the father. Experiments with animals have shown, however, that if all the code for a specific protein is contributed by either the mother or father alone, the embryo doesn't develop normally. Furthermore, the abnormalities are different, depending on which parent was the sole contributor. This research confirms something we've known for a long time—it takes both a mother and father to make a baby. But if the notion of so-called genomic imprinting proves to be true, it will clarify further the mysterious workings of the genome, including some puzzling hereditary diseases.

The daunting task of identifying all these genes and control codes may be simplified by what scientists think are great stretches of noncoding, or seemingly extra, DNA. They may regulate gene activity or offer instructions for their own replication. Recent research suggests that this extra DNA may play a role in genetic diseases. Apparently, the sequences sometimes jump out of place and crowd in between sections of genes that code for important activities. The extra code can cause the cell to ignore part of the gene's instructions, so that the wrong protein is produced.

Automated systems have taken over the task of sequencing the 3 billion base pairs in the genome. The completed road maps for human DNA would fill 500 volumes containing 1,000 pages of 1,000 words each. Technological advances in decoding DNA have enabled biologists to add 10 million base pairs to the library each year. But the real task lies in understanding what all those genes do, and scientists probably will be deep in study for hundreds of years. Although some scientists predict that, because of the rate at which technology has accelerated biological discovery, more will be learned about the cell and its genetic instructions in the next twenty years than in all the previous 200 years of research.

PROTO-ONCOGENES AND ONCOGENES

The human genome contains several dozen genes called proto-oncogenes. They are a natural part of the cell and play various roles in cell growth, but they can also be turned into oncogenes, or cancer genes, by mutations—changes within the genetic code that create a new feature. Mutations are responsible for the beneficial variety that exists in the world, as well as for many diseases.

The term oncogene is derived from the Greek word *onkos*, which means swelling or tumor. Proto-oncogenes may change into oncogenes by exposure to a virus, chemicals, radiation, or some other assault. They may be attacked by anti-oncogenes, whose job it is to counteract the influence of oncogenes on cell growth. The anti-oncogenes must be damaged, destroyed, or simply missing before the crazy cellular growth of cancer can occur. Anti-oncogenes are part of a complex and subtle defense system against cancer that is not well understood.

The prevalence of certain kinds of cancers in families may be attributed to genetic alterations in oncogenes or to tumor-suppressor genes that leave the cell vulnerable to environmental assault. There seems to be a particular genetic factor involved in ovarian, breast, colon, and skin cancers. Up to 10 percent of new cancer cases (some 100,000 annually) may be triggered by inherited genetic glitches. Although many other cancer cases seem to have familial patterns, the genetic links have not been found. Researchers suspect, however, that cancer may always be caused by some sort of damage to DNA, the blueprints that are housed in the nucleus of every cell.

Some evidence exists that certain viruses may play a role in cancer development by tampering with or mimicking the oncogene's instruction code. Retroviruses, which disrupt the cell by tricking it into producing viral DNA rather than its own, have a code sequence similar to the human oncogene. It's possible that the viral code activates the sequence of events that leads to the development of a tumor.

Some research has suggested that diet plays an important role in oncogene activity. Research with animals has shown that extremely restricted diets prevent the expression of genes that can trigger cancer. And diets high in fruits, vegetables, and cereals lower the risk, some researchers claim, by providing substances called antioxidants that protect DNA from damage and also repair it when it's broken.

POLYMERASE CHAIN REACTION (PCR)

Scientists have been able to learn much about human genetics by using a variety of sophisticated techniques, including polymerase chain reaction. PCR is a detective that snoops out a specific sequence of DNA code in a biological sample and amplifies it many times over so that it can be measured. This technique is an important development in the quest to unlock the secrets hidden within DNA. It has streamlined and accelerated the isolation and analysis of DNA, a feat that was previously accomplished only by cloning it within microbes and cells. PCR turned a months-long job into an overnight process.

PCR is useful to physicians and researchers who are investigating genetic diseases, diagnosing illnesses, analyzing specimens from crime scenes, and studying the genetic underpinnings of museum specimens. It is also a useful new tool for researchers who are mapping and sequencing the human genome.

The procedure is rather like concocting a pot of soup that contains some intelligent ingredients. The DNA to be studied is stirred together with a pinch of the nucleotides that are the chemical building blocks of the genetic code. Also tossed in is a helping of polymerase, an enzyme that assembles nucleotide building blocks into a copy of the sample DNA strand. But that can't happen without the two primers, sequences of DNA, each 12 to 20 bases long, that match the DNA before and after the region of interest in the genome.

It might even be useful to think of PCR as a pot of alphabet soup. The DNA sample under study is comparable to a sentence fragment, and the primers are analogous to the words on either end of the fragment. The nucleotides are spare letters floating around in the soup. The polymerase uses them to build a complementary copy of the sample strand, filling in between the two primers.

After the ingredients are all mixed together, the soup is heated, causing the double helix of the sample DNA to pull apart. While the soup cools, the primers take their place on the appropriate ends of the separated strands, and the polymerase extends these primers to build complementary sequences. The result is two complete double strands of the sample. Then the process is repeated many times, each complete double strand separating and being used as a template for a new strand. In a few hours, the sample DNA has been copied and recopied millions of times, yielding enough material to study.

Researchers have used PCR to observe more closely the trickery that the AIDS virus uses to fold itself into human cells and avoid elimination by the immune system. PCR has proven useful in diagnosing the presence of AIDS in infants, a task that antibody tests can't perform because of the presence of the mother's own antibodies in the infant's system. PCR can find the very small concentrations of the AIDS virus that are invisible to other tests. Researchers have also used this technique to search blood samples for the spirochete bacterium that causes Lyme disease, an illness spread by a certain kind of tick that is difficult to diagnose.

With PCR it is also possible to do evolutionary studies of crumbling museum specimens by amplifying the tiny remaining shreds of DNA and comparing them to the DNA of living descendants. Researchers who have traced the modern human to one ancient mother-of-us-all (see MITOCHONDRIAL EVE, page 90) have been quickly able to confirm their painstaking work.

PCR has also been used to find previously unknown genetic hitchhikers in the human genome, artifacts of ancient retroviruses that were simply incorporated into the human blueprint somewhere along the evolutionary road. There may be thousands of them hiding in each person's DNA, along with as many as twenty recessive diseases that sleep quietly until a sexual liaison with the wrong person awakens the potential disasters they represent.

MITOCHONDRIAL EVE

An old moral adage teaches us that we are all brothers and sisters under the skin. Now, biochemists suggest that this is literally true. By examining scraps of DNA, they claim to have found a genetic mother of all humanity—referred to as the mitochondrial Eve—who probably lived in Africa some 200,000 years ago. The research suggests that her descendants migrated throughout the world, somehow outpacing other early humans and becoming dominant.

Anthropologists disagree with the notion that modern humankind is so young. It is from their work examining bones and other fragments of ancient life that we have our understanding of the history of humanity. They claim that modern humans emerged at least a million years ago, evolving separately in various regions of the world after early ancestors had most likely migrated from Africa. The notion that some superior human upstart migrated from this genetic Eve's hometown and dead-ended Neanderthal, Peking, and Java man has caused shouting matches at scientific meetings.

The Eve hypothesis began when researchers started looking for clues to our history in human mitochondria—tiny organs that live in the cell and convert carbohydrates into energy to run the cell. Mitochondria were once free-living bacteria that, somewhere in the evolutionary journey, were recruited for service in more complex organisms that became human. Thus, the DNA contained in the mitochondria is different from that which lives in the nucleus of every cell and contains the master plans for the life of the organism. Nuclear DNA comes from both parents and is mixed up and changed with each generation. Mitochondrial DNA (mtDNA), on the other hand, comes from the female parent and passes largely unchanged from generation to generation. It is changed only by random mutations, mistakes the genes make in copying the code, that seem to occur regularly. Biochemists claim to be able to count the mutations and the time periods between them, and thus trace the beginning of modern humans.

How did they do it? First, they found 147 pregnant women who would allow researchers to use tissue from their placentas following delivery. The women's genetic ancestry represented major geographic areas of the world—Africa, the Middle East, Asia, Europe, Australia, and New Zealand. After having isolated the mtDNA from each baby and compared samples, researchers found very few differences among them. Those closest to African roots showed the most

Mitochondrial Eve

Mitochondria are the energy factories in the cell. Because they have separate DNA, it's thought that they may have been free-living organisms that formed a symbiotic relationship with early cells. Their DNA is found only in the egg and passes largely unchanged from one generation of women to the next and as a result, it is easier to trace.

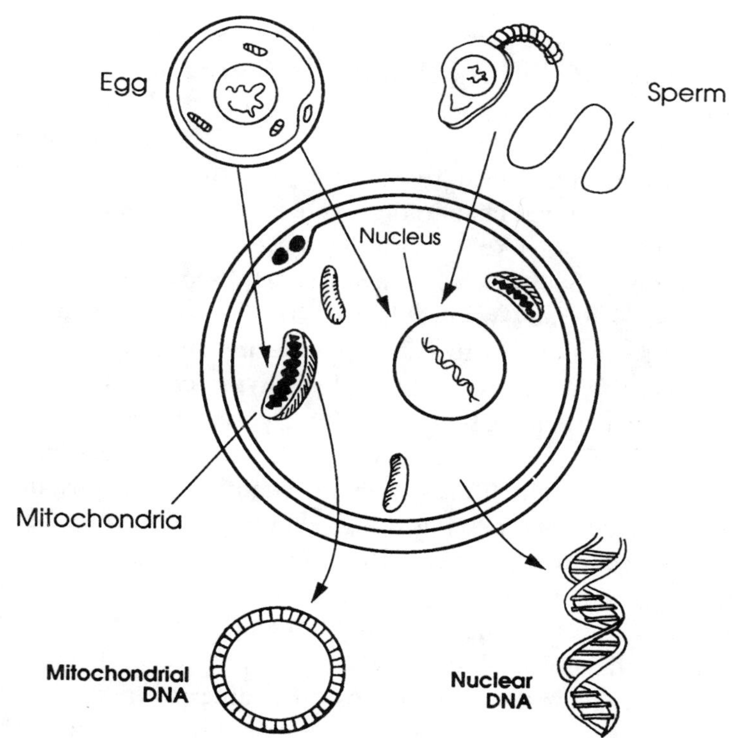

diversity, leading the researchers to believe that they had been evolving longer and thus were the line from which modern humanity descended.

Recently, a researcher analyzed the data from which these conclusions were drawn, claiming to have found statistical flaws that undermine the hypothesis that Homo sapiens began in Africa with a single mother-of-us-all. Anthropologists who have always opposed the idea are delighted, but the members of the research team who came up with the idea of African genesis continue to stand by it and are looking for new tests to support it.

Other researchers have set out to find our paternal ancestor by examining the male Y chromosome, which carries the gene that triggers male sexual development. Researchers hope to find common genetic fingerprints that will lead them back in time to Adam.

GENETIC ENGINEERING (RECOMBINANT DNA)

If the characteristics of life are determined by blocks of code in DNA, then it seems logical that adding to, subtracting from, or rearranging the code will change the expression of those characteristics. It sounds simple, like rearranging the links in a chain. But it wasn't until the early 1970s, when researchers discovered the restriction enzymes, that they were able to lift out sections of code by breaking apart the links in the chain at specific sites. Enzymes are organic catalysts, chemicals produced by a living cell that stimulate chemical reactions without being used up by the process. There are thousands of enzymes whose job is to jump-start the reactions that are necessary to sustain life. For many chemical reactions in the human body or any other living thing, there is an enzyme that fits precisely. For centuries, humans have found practical uses for the enzymes that occur in nature. Among their many applications, enzymes tenderize meat, remove stains from clothing, and are used in drugs and the diagnosis of disease.

Restriction enzymes, or nucleases, are able to recognize sequences of DNA code and sever the bonds holding them together.

Each of the four hundred or so known restriction enzymes has a specific spot where it breaks the chain. Bacteria use them as weapons against viruses, killing them by cutting up their DNA. Apparently, the human cell uses these enzymes to cut and paste genes that, in turn, produce needed proteins. For example, white blood cells may use these tools to make spare parts from which to build antibodies for the immune system.

Biochemists use restriction enzymes to cut apart DNA and combine the genes of one organism with those of another by means of ligases, other enzymes that glue sequences together. They can even design restriction enzymes that do not occur in nature, expanding the ways in which science can reorganize the blueprints of life.

It sounds like the stuff of horror movies . . . scientists breeding monsters by combining the genetic potential of wildly different creatures. Fortunately, nothing like that results from this procedure. The only remotely monstrous creatures to come out of genetic research have been bigger carp, disease- and pest-resistant tomatoes, and giant mice, which owe their size to genes from rats.

One of the most beneficial outcomes of this kind of research has been the production of hormones and other useful proteins. For example, human insulin can be produced in quantity to replace animal insulin, which has been used for years but is not suitable for all diabetics. Insulin, normally produced by the pancreas, enables cells to absorb sugar from the bloodstream for energy. Diabetics can't produce enough insulin on their own and need daily doses of the hormone. Insulin can now be produced in the lab by genetically altering E. coli bacteria, common residents of the human intestinal tract, and taking advantage of their asexual reproduction in order to clone, or copy, DNA from other organisms.

Because of their simplicity, bacteria are well suited to the production of proteins through genetic engineering: In addition to the DNA in their chromosomes, they have a separate little ring of DNA called a plasmid that is more easily manipulated. The physical differences between nuclear and plasmid DNA make it possible for scientists to separate them through centrifugation, in which centrifugal force sends larger particles out farther from the center of a spinning vessel. Once altered by genetic engineering to include extra sequences, the plasmids are easily taken up by bacteria in which

Genetic Engineering

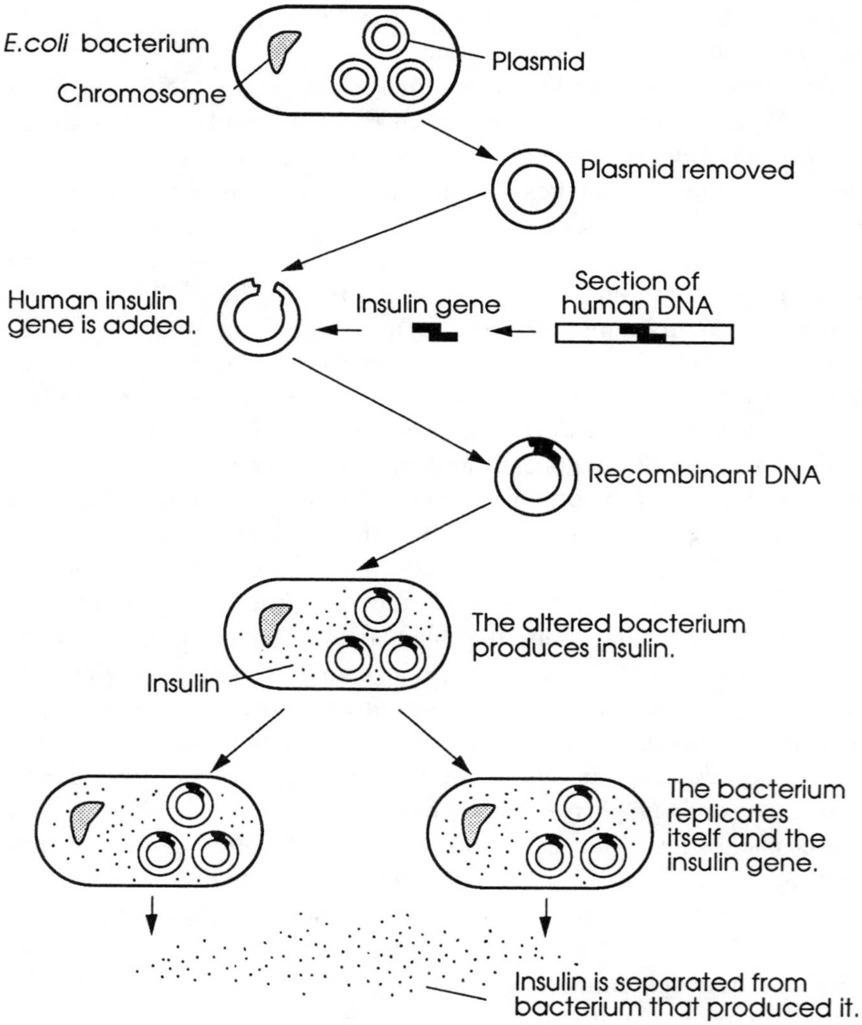

E.coli bacterium

Chromosome

Plasmid

Plasmid removed

Human insulin gene is added.

Insulin gene

Section of human DNA

Recombinant DNA

The altered bacterium produces insulin.

Insulin

The bacterium replicates itself and the insulin gene.

Insulin is separated from bacterium that produced it.

the outer wall has been weakened. The bacteria busily go about replicating themselves as if nothing had happened. They grow in tanks, replicating or cloning their own genetic information, as well as the stowaway human gene. If the human gene has been expressed in a stable fashion, it is transcribed into RNA and the RNA produces useful quantities of a vital human hormone.

Human insulin was the first genetically engineered protein to receive approval from the Food and Drug Administration. Now, other substances are being produced in bacteria fermentation tanks. For example, interferon, which the body produces naturally to inhibit the growth of some cancer cells and viruses, is being manufactured by bacteria and yeast. Interferon seems to prevent viruses from replicating, and it marks cancer cells for destruction by the warriors in the immune-system army. Scientists can even combine genes from the different kinds of naturally occurring interferon to create a substance that has the best qualities of all of them and works better on specific kinds of cancer.

Human growth hormone (HGH), previously obtained from cadavers, is now manufactured by bacteria. Children with hypopituitary dwarfism have a defective gene that renders them unable to produce this substance. The cadaver-derived HGH was implicated in the spread of a dangerous virus. Now, with the availability of genetically engineered HGH, children have a safer alternative. Tissue plasminogen activator (TPA), a substance the body produces to dissolve blood clots, is now also being manufactured in large enough quantities to treat heart-attack victims.

While bacteria have been extremely useful in the production of TPA, insulin, and other substances, some scientists are working on what they consider to be more efficient organisms for producing drugs. By injecting fertilized eggs from goats, sheep, pigs, and cattle with human genes and then implanting them in the uteruses of animal surrogate mothers, researchers have been able to produce animals that make human proteins. There are goats that produce TPA, sheep that produce an enzyme that helps treat an inherited form of emphysema, cows that produce antibacterial proteins, and pigs that make human hemoglobin, a substance in the blood that carries oxygen from the lungs to the tissues of the body and transports the carbon-dioxide waste to be exhaled.

As you may imagine, it isn't easy to insert DNA into cells. More often than not, they are decidedly stubborn about accepting it, but experiments with plant, animal, and insect tissues have yielded some interesting techniques.

Some scientists have used microscopic syringes to inject DNA into thousands of insect cells, in the hope of getting one that plays along with the idea. Others have tried the shotgun approach, in which insect embryos are blasted by DNA-covered particles. It's much quicker but, not surprisingly, most of them die. Another method involves making a kind of insect-DNA milkshake. Insect eggs are shaken up in a mixture that includes water, DNA, and a bunch of tiny silicon needles that poke holes in the eggs during the mixing. The eggs that don't die suck up the DNA, and the mature survivors bear the characteristics they acquired in the shake.

In the early 1980s, biologists developed techniques for moving genetic material from one kind of plant to another. By engineering the DNA in a kind of bacteria that infects plants, scientists have been able to create thousands of transgenic plant lines with resistance to certain insects and viruses. They can make fruits that are slow to spoil and plants that are higher in nutrients. They also remove the genes in bacteria that cause tumors and insert genes that code for desirable characteristics. Then the plant cells are infected with the bacteria, which serve as genetic transport vehicles into the cell.

Researchers use the shotgun method to blast DNA-coated particles at plant cells, too. Then the altered cells divide and regenerate into tiny plantlets and eventually full-grown plants that can pass on the transplanted genes to their offspring. Some of the crops that have been altered genetically include tobacco plants resistant to pests and tomato plants resistant to spoilage. Scientists have also created plants tolerant of herbicide applications that kill weeds. In a variation on this theme, they have equipped crop plants with a gene that codes for an enzyme that inactivates an herbicide. The weeds, of course, do not have this built-in protection.

Scientists claim that techniques to produce higher quality nutrients in plants through genetic engineering may be part of the answer to the problems of a hungry world. And plants may also be genetically altered to produce needed drugs and other chemicals. There is a downside, however.

Ever since the beginning of genetic engineering research there have been heated debates over the dangers of fooling with nature's blueprints. Critics have worried that scientists will be tempted to alter the reproductive cells of humans and make changes that will pass from generation to generation. They have also worried that some altered forms of life will escape from labs and holding tanks to upset the delicate balance in ecosystems or cause devastating diseases. To forestall such fears, scientists are trying to harness the capability of some bacteria to kill themselves. Once these bacteria have done their work, they can turn themselves off by means of a gene that acts as a switch.

Some consumers are wary of genetically engineered products and have fought their commercial use. Activists nearly stopped open-air testing of a genetically engineered bacterium that, in its natural state, has a substance in it that triggers frost on plants during near-freezing temperatures. Researchers removed from the bacterium the gene that codes for the frost-causing substance, and when they were finally able to test the altered bacteria on plants in the field, they found that the gene surgery had worked and the bacteria had not escaped from the test field as activists had feared.

Recombinant bovine growth hormone (rBGH), also known as BST or bovine somatotropin, is a protein hormone that is produced naturally by cows and is also produced in bacteria through recombinant DNA technology. It can stimulate growth in young cattle and increase milk production in cows. Researchers have been testing it since 1928. At first the hormone was collected from the pituitary glands of cows. It is now produced in quantity by *E. coli* bacteria using a method that is similar to the one used to produce human insulin.

Critics contend that the hormone, when given to cows, can pass to consumers in milk. However, the FDA has ruled that milk from cows treated with rBGH is safe. It occurs naturally in milk, and its level does not increase even when many times the suggested dosage is administered to a cow. Furthermore, researchers contend that rBGH and other proteins taken by mouth are simply digested before they have a chance to enter the blood stream. The hormone is specific to cows and not active in humans.

However, rBGH does stimulate the production of another kind of hormone in cows—insulin growth factor-1 (IGF-1)—that also is active in humans. Some researchers have expressed concern over the possibility that IGF-1 causes breast cancer in women and abnormal growth patterns in children. When ingested, IGF-1, a large protein that can't be absorbed intact, is also rendered inactivate by the digestive system. Human milk contains more IGF-1 than cow's milk, as does human saliva. We swallow more of this growth hormone in our saliva everyday than we would consume in a glass of milk from a hormone-treated cow. Nevertheless, the FDA has yet to give final approval to the commercial marketing of rBGH.

Despite critics who have tried to stop genetic engineering research and any practical applications of its discoveries, progress has continued, albeit more slowly. The controversy has also stimulated vital public discussion of its implications. And it's interesting to note that the scientists who learned how to manipulate bacterial DNA were the first to recognize the importance of controlling the technology. Concerns among the researchers themselves led to widespread discussion within the scientific community and the development of guidelines for research.

GENE THERAPY

Gene research holds the promise of tremendous gains, as well as potential risks. Of the 4,000 or so diseases caused by genetic defects, more than 1,600 are caused by a single genetic mistake. What if doctors could send tiny repair trucks into the nuclei of human cells, where workers would untangle the twisted cables of DNA, find the mistake, and fix it? People with genetic diseases like cystic fibrosis, hemophilia, polycystic kidney disease, sickle-cell anemia, and muscular dystrophy could be treated and perhaps cured.

While it is only possible in the movies to shrink a repair van and its crew down to cell size, it is possible in real life to use viruses as transport vehicles into the cell. Biochemists have learned how to convert an otherwise mischievous invader into a helpful worker.

A virus is little more than a DNA transport vehicle anyway. Barely alive, it is a protein shell that carries a payload of DNA, which it injects into the cells of whatever creature it has invaded. There it is capable of fiddling with the DNA of the host cell, interfering with its normal function, or using the environment of the cell as a kind of hatchery, where it duplicates itself many times over until the cell is destroyed. Then the duplicate viruses are set loose to increase their numbers by invading other cells.

Researchers are attempting to harness the virus's natural gene-altering abilities for medical purposes. American and French scientists, for example, have used a cold virus to carry repair genes into the lungs of rats. The virus is stripped of the genes that cause colds, and the appropriate human gene is inserted. The goal is to develop an aerosol spray to carry a viral repair truck full of spare parts into the cells of human lungs suffering from cystic fibrosis, a deadly disease that clogs its victims' lungs with a thick mucus, restricting breathing and encouraging infection. Some 30,000 Americans have cystic fibrosis, while millions more carry the defective gene that causes it.

By 1991 the National Institutes of Health (NIH) had approved five gene-therapy experiments on humans. In the first experiment to gain approval, a little girl received a gene that codes for adenosine deaminase (ADA), the lack of which seriously impaired her immune system. The gene was inserted into some of her white blood cells, which were returned to her system. The procedure worked well enough to eliminate her need for daily injections of genetically engineered ADA. She received them every few months instead.

Another approved gene-therapy experiment is the genetic engineering of liver cells from patients suffering from familial hypercholesterolemia, an inherited type of high cholesterol. Researchers used a virus to implant new genes in liver cells that remove cholesterol from the blood, and they successfully tested the technique on a breed of rabbit that suffers from high cholesterol. By removing a portion of the liver, isolating the cells, exposing them to the virus that carries the repair genes, and then injecting the altered cells into a major blood vessel feeding the liver, the researchers hope to see the patients' livers once again functioning as they should.

Viruses may someday also be used to treat atherosclerosis and other cardiovascular diseases genetically by transferring genes to cells in the walls of damaged blood vessels. These altered cells would be able to repair damage and prevent further harm. While the genes that make viruses dangerous have been removed in these experiments, some researchers are concerned that they might still cause some mischief—triggering cancer or infection. Neither has happened so far during testing.

Bacteria, very useful in biotechnology for the production of insulin and other substances, have also been recruited for use in gene therapy. In 1989 researchers at the National Cancer Institute in Bethesda, Maryland, were the first to inject a genetically engineered protein into a human with a cancerous tumor. They had treated white blood cells with interleuken-2 (IL-2), making them more powerful weapons in the immune system's arsenal. Then they needed a way to track the cells and evaluate their durability and effectiveness. So they used a special bacteria gene as a marker, splicing its genetic code for resistance to antibiotics into the souped-up white blood cells and injecting the cells into the patient. By treating samples from the patient's bloodstream with the antibiotic that the bacteria gene resisted, they were able to see if any of the treated cells had survived. After three weeks, the marker gene was still there, the cells had done their work, and the patient's tumor had shrunk by 50 percent.

The same research team has plans to genetically engineer a super cancer-fighting cell. They have been cloning the genes for IL-2 and tumor necrosis factor, which inhibits the growth of cancer cells, and intend to splice these genes into white blood cells that attack tumors. The white cells, reinjected into the bloodstream, would produce their own cancer-fighting drugs, delivering the right amounts where they are needed.

While these and other methods for treating human illness with genetic therapy show promise, it will be a long time before people can go to the doctor and be injected with genetic repair trucks. Techniques that work in mice and other mammals do not always translate to humans. Many diseases, as well as the basic characteristics of human beings, seem to be determined by a number of genes working together. Research has a long way to go before the full complexity of the human genome is understood.

ANTISENSE GENE TECHNOLOGY

Once scientists learned how to unzip the double helix and identify some of the genetic code sequences for certain functions, it was only a matter of time before they found a way to turn those codes on and off. Borrowing a trick from viruses and bacteria, they are refining a technique that essentially ties the hands of certain of the cell's code masters so that they can't do their jobs. This ability is giving researchers valuable insights into the workings of genetic codes, and it may lead some day to highly specific weapons against disease.

When a living cell needs a particular job done, the DNA coiled up inside the nucleus unzips the part of the double helix where those instructions are located. Only one side of the ladder—the "sense" strand—carries the appropriate code. Messenger RNA (mRNA) assembles itself in a mirror copy of that strand and goes out to the cytoplasm where it serves as the recipe the cell uses to make the protein needed to get the job done.

But what happens if you insert into a cell a strand of code that is flipped or reversed? Will mRNA copy it as well? The answer is yes. Researchers found out, using genes from simple organisms, that it's possible to fool Mother Nature in this way. By either injecting the reversed mRNA strand or manipulating the DNA in the cell so that its own mRNA copies the wrong side of the ladder, they succeeded in fooling the cell's system for copying and reading code. When the mRNA copies the wrong sequence, it delivers the wrong message to the ribosomes, which act as copying work stations in the cell.

It is also possible to create a situation in which a reversed mRNA bonds with the original mRNA strand, its opposite, thus dead-ending them both. The unhappy couple is broken down by enzymes, and not able to bind to the ribosome to produce a protein.

Bacteria and viruses use antisense RNA naturally to regulate gene activity, and researchers have borrowed this technique to interfere with the production of proteins that determine softness in ripening fruits and color in flowers. They have also managed to interfere with a virus that infects tobacco plants, a leukemia virus in mice, and human cancer cells grown in culture.

Biotechnologists in California have engineered several generations of tomatoes whose cells contain a flipped version of the messenger

RNA from the gene that codes for the enzyme that makes ripe tomatoes mushy, thus blocking that enzyme from destroying the tomatoes. Growers would be able to leave these tomatoes to ripen longer on the vine, making them tastier and better able to travel.

By inserting cloned gene sequences for the production of certain pigments, researchers have changed the color of flowers. It is also possible for them to turn off a gene that is already part of the plant's DNA by injecting the cells with flipped copies of a pigment code. The cell makes mRNA from the antisense code and prevents the plant from producing pigment.

The ability to select plant characteristics through antisense genetics has great economic potential, but it's the possibility of turning off the genes that cause human disease that could have the biggest impact on our lives. For example, if doctors could inject cancer patients with antisense RNA, they could turn off or rehabilitate cancer genes in a very specific way without harming non-cancer cells. Current cancer therapies kill healthy cells along with the diseased ones, making the patient even sicker for a while.

Retroviruses, like human immunodeficiency virus (HIV), which causes Acquired Immunodeficiency Syndrome (AIDS), use similar trickery to get human cells to duplicate viral RNA and create new copies that go on to invade other cells (see VIRUSES, RETROVIRUSES, AND AIDS, page 107). The virus uses an enzyme called reverse transcriptase to make a complementary DNA copy of its own RNA, and then fools the human cell's DNA into incorporating the sequence into its own code strand. Having commandeered the human cell's control center, the viral genes instruct the cell to make duplicate copies of the virus.

Scientists have also succeeded in using antisense RNA to dead-end a leukemia virus in mice. They inject mouse embryos with antisense RNA that blocks the code that puts new copies of the virus's genes into their protein shells. When the host cell releases its newly produced viruses, they are just empty shells incapable of infecting other cells. As a result of being equipped with antisense genes, the grown mice are able to resist infection by the leukemia virus.

While prenatal treatments such as the one just described may never be used on humans, doctors may someday have antisense drugs for genetic diseases and for the treatment of cancer and viral infections.

CLONE

Despite the horror stories and jokes about powerful people having themselves cloned, this technique has been limited to the study of the cell, the duplication of DNA, and the asexual reproduction of many kinds of plants, as well as the copying of some frogs, fish, and mammals.

Cloning occurs frequently in the plant world, and for thousands of years humans have known about and used the plant's ability to reproduce both sexually and asexually. Instead of waiting for the time-consuming process of pollination, growth, and seed production, a grower can take a cutting from a mature plant; its cells will soon differentiate into all the parts that will allow it to live on its own.

For example, every time you give a cutting of your favorite houseplant to a friend, you begin a process that results in a clone, or identical genetic copy of the original plant. When gardeners plant chunks of potato, called seed potatoes, every spring, they are growing clones of some other potato plant that had desirable qualities. Roses, apples, and many other plants are routinely cloned by inserting a twig or bud from a closely related plant into a slit on a host plant in order to achieve the best fruit or flower possible. In fact, the term *clone* comes from a Greek word that means twig or branch.

Today, scientists are able to clone complete plants by using single cells. Many plant cells that are not very specialized are totipotent, meaning they are undifferentiated, or contain and are able to use all the genes necessary to produce an entire plant. By contrast, a differentiated cell is specialized to perform only certain functions and cannot give rise to a complete organism. In other words, you can clone a mature carrot using one of its cells, but you can't clone a grown woman using just any of her cells.

It is possible, however, to clone some of the genetic instructions in the human cell. By splicing human DNA into that of bacteria or viruses, researchers have developed ways to clone the genetic instructions for important human proteins like insulin, interferon, and human growth factor. When the microbes replicate, or clone their own genetic instructions, they also reproduce their human stowaway genes. Huge vats of human protein are then separated from their microbial producers and made available to people who need them.

The ability to clone larger animals has allowed livestock breeders to preserve and amplify the championship lines of prize bulls and cows, producing farm animals with the cookie-cutter sameness and quality of modern factory products. This process is achieved by artificially inseminating a cow with semen from a champion bull and then removing the embryos, each of which has already divided into a number of cells. Each cell nucleus contains the genetic plans for an entire beast and so, when the nuclei of all the embryonic cells are transplanted into eggs from ordinary cows and implanted in their wombs, the surrogate mothers give birth to identical champion calves. Zoos have successfully used similar cloning technology to produce new members of endangered species, which are born to surrogate ordinary mothers.

It is the scientists' ability to clone mammals that makes some people nervous. If they can clone mice, sheep, and cattle, how difficult would it be to clone human beings? In fact, not difficult at all. The technology exists.

Medicine's ability to manipulate the human reproductive process through artificial insemination and *in vitro* (which means "in a glass dish") fertilization anticipates, in some minds, the cloning of human beings.

In theory, it would be possible to produce a human clone by removing the nucleus of a certain kind of reproductive cell that has a full complement of genetic material and exchanging it for the nucleus of an egg cell. In 1979, Landrum B. Shettles reported in the *American Journal of Obstetrics and Gynecology* that he had done just that. Three such eggs started to divide, but they were not implanted into a human uterus and allowed to develop.

However, even though it may be possible to transfer human embryonic cell nuclei into other human eggs, it is not likely to happen for ethical and economic reasons. No responsible scientist would take on such an experiment, nor would any funding agency, such as the federal government, support a researcher intent upon manipulating human beings in that way. Such work would require the support of a sophisticated laboratory, plenty of money, and a distinct lack of scruples.

At this writing, it is not possible to make clones from the DNA of mature humans. So we are not likely to see little copies of Albert Einstein or Abraham Lincoln any day soon. (Samples do exist of Einstein's brain tissue, as does blood from the shirt sleeves of the surgeon who treated the assassinated president.)

It's not possible in this case, because there isn't a complete copy of either man's DNA available. Even if there were, scientists don't understand enough about the human genome to make a whole human from DNA. Furthermore, human cloning won't work because proteins in the mature cell restrict the developmental potential of DNA, making it an unsuitable recipe for a whole new human being. Scientists may someday figure out how to reverse these restrictions, but to what end? The mature adult is more than the sum of his or her parts. Experience, knowledge, and memory cannot be cloned.

Nevertheless, ethicists tell us that the pace at which genetic research and engineering are advancing demands that we carefully discuss such issues and decide how we will use this knowledge when it becomes available.

DNA FINGERPRINT

The techniques that are unraveling the secrets of disease and inheritance are also being used to investigate crimes. DNA fingerprinting is based on the notion that an individual's genetic blueprint is unique. Forensic scientists believe that these differences can be used to determine the identity of a person who is suspected of a crime.

This "fingerprint," taken from tissue, hair, blood, or semen found on a victim or at the scene of a crime, is compared to a suspect's DNA. If it matches, and the evidence is permitted by the court, it can mean the difference between imprisonment and freedom for the suspect.

Making that incriminating match involves a complicated process that critics claim is unreliable, however. Samples taken from a crime scene are often small, preventing lab technicians from doing the kind of retesting that is needed to verify results. And, unlike the specimens used in molecular biology research labs where the process was developed, samples from the scenes of violent crimes are often degraded by age and contamination. Nevertheless, the Federal Bureau of Investigation has embraced the use of DNA fingerprinting, and a 1992 Yale University study has found that the likelihood of a chance match is no greater than one in a million.

This seems remarkable when you consider that as much as 99 percent of human DNA is identical from one individual to the next. After all, most of us have the same basic body parts; it's how they're put together that makes us different. In relatively closed communities and tightly knit families there may be very similar code sequences, but only identical twins have precisely matching blueprints.

Hidden within the vast sameness of the human genome are little stretches of repetitive code that cluster here and there in highly individualistic patterns. No one knows what they do, but British molecular biologist Alec Jeffreys at the University of Leicester discovered in 1985 that these repetitions could be used as identity tags for individuals. For example, one person's DNA may contain clusters of ten repeats of a certain code, while another person may have fifteen repeats in a cluster. DNA fingerprinting isolates the segments that contain clusters and compares them.

In the lab, DNA is extracted from the sample and mixed with restriction enzymes that cut the double strand at specific points, leaving millions of tiny pieces. (See GENETIC ENGINEERING, page 92.) These fragments, called restriction fragment length polymorphisms (RFLP), are placed at one end of a piece of gel to which a negative electrode is attached. When a current is turned on, a positive electrode at the other end of the gel attracts the negatively charged DNA fragments. Like microscopic worms on a slab of Jell-O, these fragments wriggle toward the positive electrode until the current is turned off. Not surprisingly, the shorter, lighter fragments win the race. Meanwhile, all the pieces have sorted themselves into a pattern according to size on the gel.

These fragments are transferred onto a special pad that is then saturated with radioactive DNA probes: specially created lengths of DNA that bind to a specific sequence of bases on or near the clusters of code repetitions. When the pad is exposed to X-rays, the radioactive probes show up on film as a pattern of fuzzy bands that, in theory, are as distinctive as the patterns on our fingertips.

In addition to the less than ideal condition of the sample, the delicate complexity of gel electrophoresis, as this technique is called, can lead to false results, critics contend. The DNA from other organic substances at the scene can contaminate the sample. Age and other factors can break the RFLPs at the wrong places, and sometimes overambitious RFLPs race ahead during the swim through the gel, causing what is known as band shifting. In this case, the bands that are visible on an X-ray film will appear to be from different sources when they are actually matches. Experts have suggested that band shifting may occur as much as 30 percent of the time. The possibility of serious error is compounded, critics continue, by the willingness of jurors to accept a "scientific" solution to a criminal investigation, without questioning the accuracy of its findings.

DNA fingerprinting has been used in hundreds of criminal investigations, but it is also used in paternity cases and for tracing the inheritance of genetic diseases in families. The technique has been helpful in identifying human skeletal remains and classifying museum specimens of birds that are more than a hundred years old. Scientists hope to analyze the genetic makeup of extinct species and compare creatures from the past with those of today.

VIRUSES, RETROVIRUSES, AND AIDS

Viruses are tiny protein-wrapped packets of genetic information, a hundred times smaller than a bacterium. They are the smallest known infectious agents in the world, ranging in size from one- to ten-millionths of an inch thick.

Not quite alive, viruses are incapable of the functions that characterize living things. If they once were truly alive, as some scientists suspect, they seem long ago to have lost the ability to eat, metabolize, or reproduce on their own. In fact, they don't show any signs of life until they get inside a living cell. Then they use the worst kind of trickery to replicate themselves and cause mischief.

Inside the virus's protein shell is a scrap of genetic material—the genes that contain the instructions for the virus's duplication. By a variety of methods, viruses insert their own genetic material into the cell. In some cases, the cell welcomes the virus inside, mistakenly identifying one of the proteins in its outer shell as something friendly, like a needed chemical delivery. Once inside, however, the protein shell breaks apart and the virus's genetic material gets busy using the resources of the cell to replicate itself. Then the crowd of new viruses is released, often destroying the cell before marching off to make new conquests.

Some viruses, like the herpes viruses that cause chickenpox, shingles, genital herpes, cold sores, mononucleosis, and congenital defects, are able to invade cells and then hide, causing chronic or recurring episodes of illness. Viruses have also been implicated in certain cancers. By tinkering with the cell's DNA they can cause mutations in certain genes, called oncogenes. Once they are turned on, oncogenes cause the wild cellular division that leads to tumors. And other viruses hide out for years before they finally make their host sick.

The immune system normally has strategies for dealing with such invaders, quickly capturing and destroying the culprits. For example, with the common cold, which is caused by viruses, the immune system's speedy response causes a few days of misery until it finally gets the upper hand. But some viruses—like the ones that cause rabies and poliomyelitis (polio)—act so quickly to invade and destroy cells, or induce them to produce toxins, that there isn't time to fight.

Antibiotics, which inhibit or destroy the life functions of cells or microorganisms, are useless against a virus, something that isn't even alive. It's difficult to create drugs that will kill viruses without

Viral Replication

Viruses invade living cells and use the resources of those cells in order to replicate themselves. The host cell is often killed by the process, and it is this destruction and the immune response to the presence of the virus that makes us feel sick.

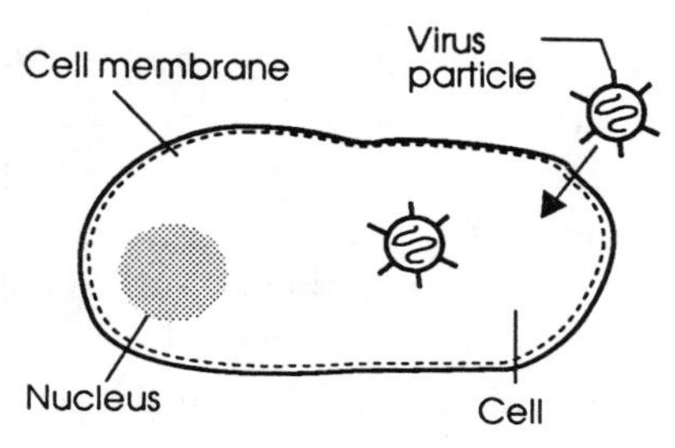

Virus attaches itself to the cell wall and injects itself or just its genetic material into the cell.

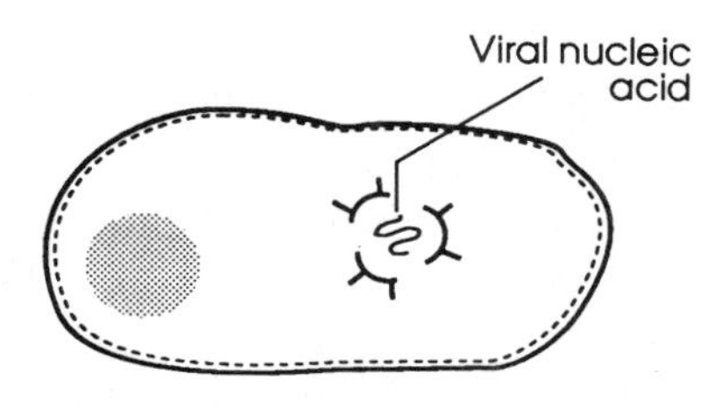

The virus's protein shell breaks apart, releasing its genes.

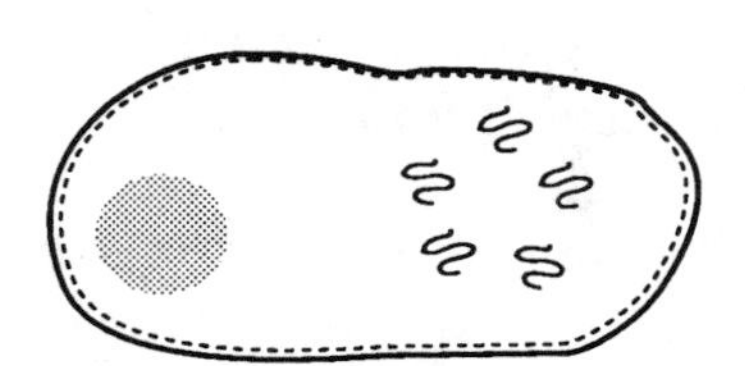

The viral genes replicated themselves using the resources of the cell.

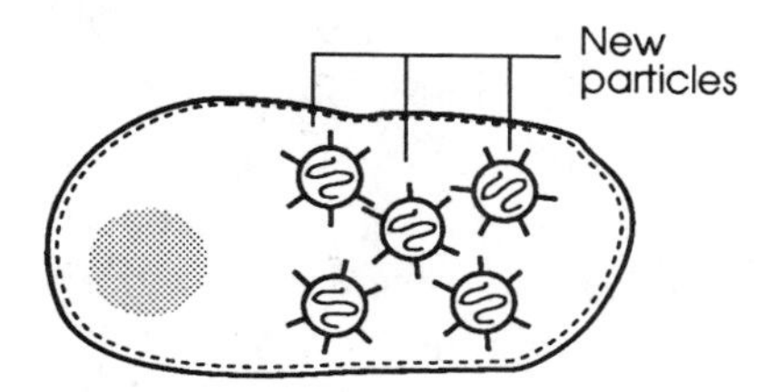

The new genes instruct for and build their own new protein shells.

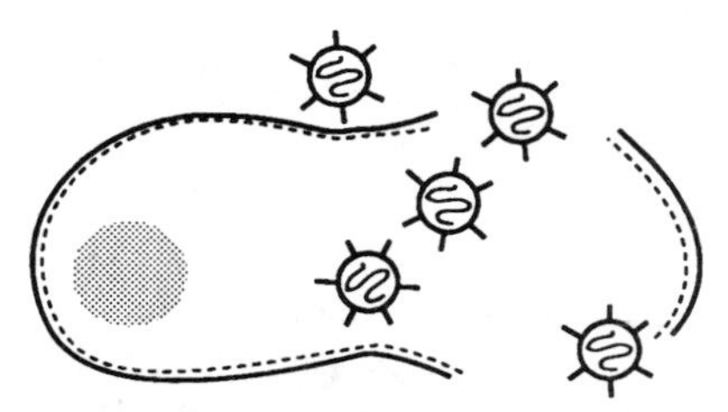

The new viruses destroy the cell and escape to invade other cells.

also killing the cells they inhabit. Those drugs that succeed in controlling viruses do so by interfering with their ability to invade the cell or replicate within it. One such drug is interferon, a natural substance that virus-infected cells produce and that researchers have learned to make through genetic engineering.

The best defense against viruses has proven to be vaccines. The most notable successes include those vaccines that prevent smallpox, poliomyelitis, measles, mumps, rubella, hepatitis B, and rabies. A worldwide vaccination program has virtually wiped out smallpox, a deadly disease that is caused by the variola virus. The world's last captive specimens of this ancient killer have been isolated in a freezer at the Centers for Disease Control in Atlanta and were scheduled for destruction in 1993 because of concern over the possibility the pathogens could be released somehow and become a danger once again.

Sometimes previously unidentified diseases seem to pop up out of nowhere. Take the case of Acquired Immunodeficiency Syndrome (AIDS), which researchers are certain evolved in Africa. The virus that causes it—human immunodeficiency virus (HIV)—may have been slowly working its way through the populace for a hundred years or more, only recently making itself obvious. The first known case of AIDS was recognized, in hindsight, through the medical records of a sailor who had died of a mysterious illness in the 1950s.

Recent research indicates that the world's oceans may serve as holding tanks for a dizzying array of viruses that cause trouble for marine life and also wait, perhaps indefinitely, for an opportunity to invade new victims. Some find their way into the oceans in sewage effluents and return to land in the bodies of the shellfish and other sea animals we eat.

Besides being crafty pirates good at hiding and wearing disguises, viruses are also changelings. They are able to mutate so quickly that the human immune system, which creates antibodies specific to each invader, is often unable to keep up. This is one of the impediments to finding a cure for AIDS. The virus seems to change too fast for the immune system or medical research to pin it down.

It's not all bad news, however. Some of the characteristics that make viruses troublesome are useful in genetic engineering. By splicing good genes into a virus's own genetic code, researchers have used them as vectors or transmitters of genetic information that can invade diseased cells, carrying repair parts in with them. (See GENETIC ENGINEERING, page 92.)

Viruses may also serve as the lesser of two evils. For example, they have been used to attack and disable immune-system cells in mice suffering from diabetes. In this case, the immune system has become destructive, attacking the insulin-producing cells of the pancreas. A mild chronic viral infection seems an acceptable trade-off for the prevention of diabetes.

Retroviruses, so named because of their retro or backwards invasion strategy, have qualities that make them especially useful tools in genetic engineering. Most viruses contain a scrap of DNA that is just sufficient to take over the machinery of the cell for its own purposes, using the material in the cell to make new viruses.

Retroviruses, on the other hand, trick a cell into doing the job for them. They contain RNA, which uses an enzyme called reverse transcriptase to build a DNA copy of itself and insert it into the cell's own nuclear DNA. Then the cell becomes the agent of its own destruction, making new RNA and its companion enzyme in order to create more viruses that will not only kill the cell but go on to invade others. Through genetic engineering, however, the virus can be engineered to carry the instructions for a good gene into the cell, manufacture it, and insert it into the cell's master plan.

HIV is a retrovirus, and researchers have learned that it cleverly invades immune-system cells, the very cells that are meant to protect the body from invaders. T4 or T-helper lymphocytes, the generals in the immune-system army, have receptors on their surface that allow them to hook onto other cells and communicate with them. These generals spit out chemical messages that direct their soldiers—including the mighty macrophage, a cell that wanders around and engulfs invaders and dead red-blood cells—to search for and destroy the enemy. They also stimulate another group of cells to make antibodies to the invader so that the immune system will recognize a particular enemy the next time it tries to attack.

HIV's strategy for disabling the immune system army is to take out the generals and break down the chain of command. Its secret weapon is a surface protein that allows it to hook onto the T4's receptor without being recognized as foreign. Using the front door, as it were, HIV marches right in and kills the general.

Other immune cells with the same kind of receptor are equally vulnerable. Macrophages are hardy warriors that are immobilized but not killed by the virus. The virus replicates itself inside the macrophage, which becomes a warehouse full of hundreds of virus particles, keeping them out of circulation for a time. As a result, antibody tests sometimes are not able to detect the presence of the virus, even though the person has already been infected. However, once the virus is set loose in the bloodstream, it can travel throughout the body in a matter of minutes.

The World Health Organization reported in 1992 that the AIDS virus is spreading faster than ever, particularly in developing nations—in Africa, India and Southeast Asia—which account for 90 percent of new cases.

HIV seems to be transmitted mainly by blood and semen, although it has been isolated in other body fluids, including saliva, tears, and breast milk. Infection has occurred primarily through sexual contact, blood transfusion, needle sharing, and childbearing, in which an infected mother transmits the disease to her fetus. It is never spread among people who have only casual contact, like hugging, shaking hands, using the same bathrooms, or eating and drinking with the same dishes. Officials assure the public that improvements in tests and testing procedures have made the blood supply in the United States as safe as possible from HIV infection.

Most new cases of infections are being acquired through heterosexual activity. In the United States, where AIDS cases had been reported predominately among homosexual males, the balance has tipped toward intravenous drug users, the poor, and minorities. Infections contracted through heterosexual activity are rising. Mathematical models predict a pattern in which the rate of infection will increase rapidly over time, unless medical researchers find an answer.

Despite years of research, no one has discovered a cure for AIDS or developed a vaccine against HIV. Studies have shown that experimental drugs such as zidovudine (also known as AZT) delay the onset of disease symptoms but do not prolong life. Recently, researchers succeeded in developing a vaccine against a similar virus that attacks monkeys. It was derived through genetic engineering from the protein shell for the simian immunodeficiency virus. This vaccine offers hope that a similar weapon against AIDS may someday be developed.

MONOCLONAL ANTIBODIES

Monoclonal antibodies are produced by specially created animal cells called hybridomas. Researchers create hybridomas in the laboratory by fusing together spleen cells called lymphocytes and cancer cells called myelomas. Why on earth, you ask, would we need such a thing?

It may surprise you to learn that monoclonal antibodies are now available at the nearest drugstore. They are commonly used in pregnancy testing products and other kits you can use at home to detect the presence of proteins in body fluids. These hybrid cells have also proven extremely useful in other areas of life besides medicine. But to understand how monoclonal antibodies work, it's necessary to talk first about antibodies.

There is always a war going on inside living things. Bacteria, viruses, renegade cells, and other foreign things called antigens are trying to get a foothold. The soldiers in this defense system are cells whose job it is to protect the organism from such threats. Standing watch are unimposing little white-blood cells called lymphocytes, which wander quietly around the body. When they sense the presence of something strange, however, lymphocytes experience a powerful transformation. They swell up and divide, giving rise to a family of warrior cells. Some of these cells produce antibodies, the special forces that prepare the battlefield, call for reinforcements and lead the attack.

Antibodies are molecules that attach themselves to foreign substances, and like a radio transmitter on a ship's hull, they locate the target for the big guns that will blow the enemy out of the water. Sometimes, after binding with an invader, antibodies act like time bombs, setting off a sequence of events that rupture the hull or membrane of the invader and sink the ship.

Each antibody molecule is constructed by its cell to bind specifically with the antigen that is present. The antibody attracts other cells, which also bind to the group, marking and entrapping the invader. Then a fearsome war machine, the macrophage (its name comes from the Greek words for "big eater"), engulfs the enemy and destroys it.

Meanwhile, still other cells retain the memory of the invader. The next time it tries to trespass, the defense system will be ready for it. Based on this capability of the cell, vaccines, using weak or dead viruses, have been developed to stimulate the immune system to produce antibodies and lock in the memory of an invader like polio, so that the body is ready to defeat it before it gets started.

Monoclonal antibodies are designed to function much like naturally occurring antibodies. They are produced by injecting an antigen into a mouse, whose immune system creates antibodies specific to the invader. The lymphocytes that cling to the antigen are then removed from the mouse's spleen and fused with cancer cells that are capable of replicating themselves indefinitely. The resulting hybrid cell combines the useful traits of both—the ability to produce antibodies to a specific substance and immortality. Once the hybridoma is created, it is isolated and cloned. Hence the name monoclonal, which means cloned from a single cell. This process produces an infinite supply of that one hybridoma.

Unfortunately, the human immune system eventually figures out that mouse proteins have wandered into its territory and begin to attack the monoclonal antibodies. Researchers have been working on combining key human genes with the mouse material so that the body will recognize the concoction as human, not mouse, in origin and therefore not attack it. They have also recently been able to create mice with human immune systems. This was accomplished by wiping out the mouse's bone marrow, where immune system cells are generated, and replacing it with human bone marrow. These mice will be able to produce human antibodies and will also be used to test drugs and vaccines.

Human monoclonal antibodies are used to fight infections caused by common bacteria that get out of control when the immune system is weak or when accidents or surgery allow microbes into the bloodstream. Some bacterial invaders ooze a powerful toxin that can cause septic shock, resulting in massive system failure and death. Although antibiotics can kill the invading bacteria, the toxins continue to cause mischief. Monoclonal antibodies go after the toxins and deactivate them.

Monoclonal antibodies are now used to type blood; to identify bacteria, viruses, and other substances in the bloodstream; and to suppress rejection of kidney transplants. One type, myoscint, is used in a diagnostic test that can assess the effects of heart disease, the number-one killer of adults in the United States. Myoscint binds with myosin, a protein leaked by dying heart cells. It is tagged with a radioactive isotope (an element that emits radiation detectable with X-ray monitors) which glows yellow when viewed by a nuclear imaging machine, allowing doctors to locate and assess the damage done by a heart attack.

Another monoclonal antibody tagged with a radioactive isotope is used to find blood clots that are sneaking around in the body, threatening to damage the heart, lungs, or brain. In addition to chemical cancer killers, radioactive isotopes can deliver a highly localized zap of deadly radiation to cancer cells as the monoclonal antibody binds to them. A monoclonal antibody that would detect plaque in blood vessels is also being developed. It would detect small amounts of plaque and allow doctors and patients to take action before the blood vessels clog, thereby preventing heart disease.

Researchers have not confined themselves to medical applications for monoclonal antibodies. For example, a Japanese electronics company has successfully used monoclonal antibodies in a bomb-detection device that takes as little as two minutes to detect the nitrogen-based compounds in gas given off by explosives.

BLOOD SUBSTITUTES

Whereas white cells are the soldiers of the immune system, platelets are the structural engineers of the blood, rushing to the site of an injury to begin building clots.

Blood is the human body's lifeline to the outside world. It carries oxygen and food to the cells and hauls away carbon dioxide, the waste product of metabolism. It regulates body temperature, transports hormones and hundreds of other proteins from cell to cell, and mobilizes the foot soldiers in the body's immune system. We each carry around five or six quarts of the stuff, accounting for about 10 percent of our body weight. And though scientists have been trying to create a convenient universal substitute, so far nothing will do like the real thing.

In addition to platelets, or thrombocytes, the main components of blood are plasma (a yellowish liquid that accounts for about 60 percent of the volume), erythrocytes, or red blood cells, and leukocytes, or white blood cells. Plasma, which is mostly water, transports some of the hormones, antibodies, and wastes that circulate in the blood. It's the red blood cell, with its ability to grasp and carry oxygen to the tissues, that is the most critical part of a lifesaving transfusion.

There are plenty of reasons to search for a safe and effective blood substitute. Whole blood is extremely perishable, lasting little more than a month in refrigeration. A mass-produced, long-lasting blood replacement that didn't need refrigeration would be especially helpful to soldiers wounded on the battlefield, as well as to accident victims everywhere.

The human body also is very picky about the kind of blood it uses. Each individual needs one of four specific types—A, B, AB, or O, either Rh-positive or Rh-negative. Otherwise, the body will violently reject the transfusion, a reaction that can easily be fatal. Scientists would like to find a universal substitute that could be used without worries about rejection.

Any successful blood substitute must somehow replace hemoglobin, the tiny, iron-rich molecule in the red blood cell that actually binds with oxygen. There can be as many as 5 million red cells in a drop of blood, and each cell contains 300 million or more hemoglobin molecules, each of which binds with the oxygen molecules that are drawn into the body by the lungs. It's easy to see why human blood has an incredible capacity for moving oxygen.

Experiments with blood substitutes have concentrated on isolating and using pure human hemoglobin or creating substances that mimic hemoglobin's abilities. Scientists have been working for more than fifty years and have yet to crack all the secrets of the hemoglobin molecule.

Pure hemoglobin has a room-temperature shelf life of six months or more, but when injected into the blood stream it is a short-lived and miserly chemical that grabs onto oxygen but won't let go. It also has the unpleasant habit of breaking into two pieces that are useless and, what's more, harmful. The kidneys filter the fragments out of the blood, but, in the process, are somehow hurt by them. Experts disagree over the reason, attributing the problem either to stray pieces of blood-cell membrane or to endotoxins—poisons produced by certain bacteria when they disintegrate.

Researchers have tried to overcome the problems with hemoglobin by chemically altering the molecule to make it sturdier and more willing to let go of oxygen. They have spliced the human genes that code for a desirable form of hemoglobin into the DNA of yeast or bacteria. Huge vats of the altered bacteria, swimming in nutrients, turn bloodred overnight as the bacteria multiply. After the bacteria have been broken open, all the unwanted proteins are filtered out, leaving what researchers hope is a pure form of human hemoglobin.

A number of scientists have genetically engineered larger animals that can produce human hemoglobin in their own red-blood cells. This is accomplished by injecting human genes into the newly fertilized eggs of the animals. One group has produced transgenic pigs that can make human hemoglobin in about 15 percent of their red-blood cells. Pigs were chosen for the job because of the same qualities that endear them to farmers—they have big litters relatively quickly. They also have plentiful blood supplies, which they can share without suffering.

Side effects continue to plague tests of animal- and bacteria-produced hemoglobin on people. Critics worry that some animal pathogen (a disease-carrying organism) or bacterial endotoxin could linger in these genetically engineered hemoglobins. Some speculate that there may be something toxic about hemoglobin itself when it is not encased by the red blood cell.

The alternative may be a kind of artificial blood substitute such as fluorocarbons, organic molecules that bind with oxygen even more efficiently than hemoglobin. Unfortunately, they also have produced unpleasant side effects in human testing. But such chemicals, called "white blood," have been approved in Japan for use during surgery and in the United States by the Food and Drug Administration for use during balloon angioplasty—a procedure in which a small balloon is inflated inside a coronary artery to flatten deposits and open up the vessel.

Although human blood is an efficient transporter of useful substances, it can also harbor and circulate diseases. Worries about contamination of the national blood supply with the AIDS virus, hepatitis, and other viruses have intensified the search for a safe substitute. Physicians are now more cautious about transfusions because of these worries, using the minimum amount of transfused blood after accidents or during surgery. The nation's blood supply is now considered safer than ever because of more stringent screening procedures. However, even the number of donations has decreased because of public fear about disease, even though experts assure us that there is no danger of contracting a disease by *donating* blood.

The U.S. Army is particularly interested in the development of safe blood substitutes. The race to produce such a product is also big business. In the United States alone, about 8 million units of blood are used for transfusions each year. Experts say that number would rise if a reliable substitute were developed, making the market even bigger. Meanwhile, drug companies are investing millions of dollars in a product that may be years from perfection.

CULTURED AND SYNTHETIC SKIN

Hold this page between your fingers. A layer of cells about that thick, called the epidermis, is all that stands between you and a pretty cruel world. It is the top layer of the largest organ in the human body, the skin—a busy, multilayered defense system that keeps the world out and you in. It regulates body temperature, sheds waste products, produces chemicals like vitamin D, senses the world, and wards off invaders. You cannot live without it.

Serious burns that destroy even a small percentage of the skin are dangerous because skin cells can't regenerate fast enough to keep fluid in and bacteria out. Researchers have devised ways to

Skin

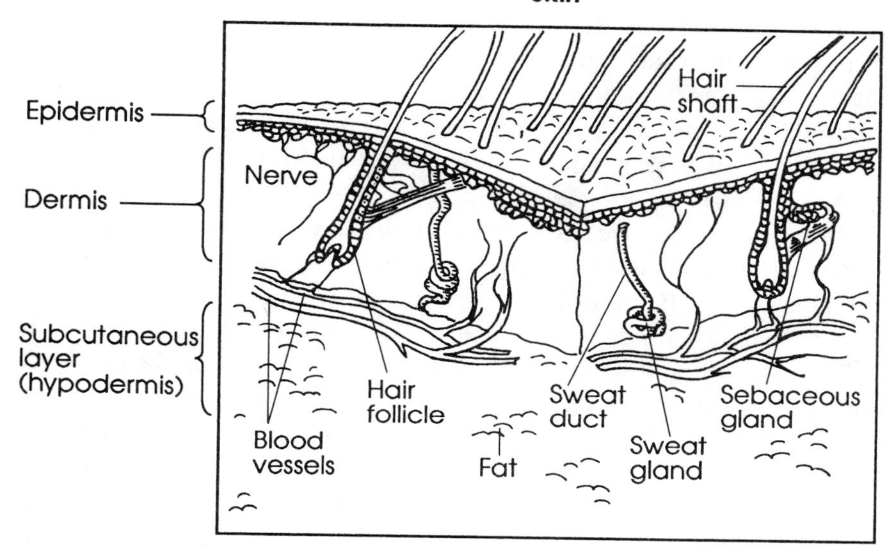

cover wounded tissue, using grafted skin from the patient, and from animals, human cadavers, and the test tube. But to understand how these techniques save lives, it's necessary to look first at how human skin works.

Think of skin as a kind of obedient society of cells, in which each member works carefully and predictably at its job. The cells in the outer layer, the epidermis, form the first wall of defense. They are dead bricks of protein bound together by a kind of fatty mortar. Farther down is a layer of constantly dividing cells, the crew that is always replenishing the supply of bricks for this protective wall. These keratinocytes, named for the tough, stringy protein, keratin, from which they're made, are programmed to find their way to the outer layer and promptly die, at which time they do their best work protecting the body. The layer that produces keratinocytes also provides melanocytes, pigment producers that send little packets of color out to tan the surface cells and protect them from the sun.

Under the epidermis lies the dermis, a thick network of protein fibers, called collagen, which is produced by fibroblasts. These cells are part of a network that includes thousands of nerve endings, blood vessels, sweat and oil glands, and hair follicles. The capillaries in these blood vessels regulate body temperature by swelling with blood to radiate heat away from the body or clamp down to conserve warmth. Some two thousand miles of tightly coiled sweat ducts regulate body temperature by flooding the surface with a slightly acidic and oily liquid that repels bacteria and cools the skin as it evaporates. The dermis, along with a layer of fat cells underneath it, acts as insulation and padding over the muscles and internal organs.

It takes several weeks for the keratinocyte, the major building block of skin, to develop, find its way to the outer wall, stand its post, and then flake away. It's only when the skin has suffered an injury that this mild-mannered little cell and its fellow-citizen cells stir into action and realize their full potential. They have a powerful ability to mend their society.

An injury sets in motion a dramatic inflammatory response. Special cells influence the capillaries to flush the area with blood and other fluids, causing inflammation and swelling. Then, wave after wave of white blood cells scour the area, attacking microbes that have slipped in through the broken barrier. Meanwhile, fibroblasts start making scar tissue, which will draw the wound together and make it a smaller breach for the now wildly dividing keratinocytes to fill.

Sometimes the wound is just too great, and skin cells cannot grow fast enough to stop deadly infection and fluid loss from winning the battle. There are some 100 billion bacteria swarming over normal skin at any one time. Antibiotics are used until the swelling goes down, but the best defense is quickly covering a burn wound. But with what? It depends upon the seriousness of the wound.

A thin layer of skin taken from another part of the patient's body is the best replacement. It is cut into a meshlike sheet that can be stretched to cover more area, and cells migrate to fill the gaps in the mesh. In the case of extensive burns, however, there isn't enough good skin left. Until recently, people didn't survive such burns.

Now, some sort of temporary skin replacement is used to buy time while surgeons take consecutive grafts from the remaining good skin, prepare cultures, or acquire synthetic skins. Cadaver skin, pig skin, and synthetic skins are used temporarily to cover the wound and encourage fibroblasts and blood vessels to create new dermis.

Human skin is the most immunogenic, or hostile to strange tissue, of all the body's organs. However, a burn may suppress the immune system for as many as six weeks, a condition that can be prolonged by drugs, if necessary. Cadaver skin will work as a temporary covering for several weeks, and blood vessels may even begin to grow into it. But once that happens, unless the immune system is suppressed by drugs, the skin will be rejected. Pig skin is more likely to be rejected by the immune system and is used only as a biologic dressing that will last only a few days.

Cultured skin begins as tiny fragments salvaged from an unburned spot on the patient's body. A shred of epidermis is broken up into individual keratinocytes, which are placed in a nutrient-rich solution and allowed to grow. Fibroblasts are included in the mix because they produce fluids that help keratinocytes grow. These cells will gather together in a dish just as they do on a wound, forming an intact sheet of epidermis. The new skin looks like a piece of wet tissue and is just as fragile. In about a month, postage stamp–size patches of new skin are available to cover a thousand times the area of the original piece. This new skin is not capable of growing new hair follicles or sweat glands, though.

Researchers have discovered that skin cultured from other sources works as well as that grown from a patient's own cells. Apparently, keratinocytes produce chemicals that stimulate other keratinocytes to grow, no matter whose they are. Ultimately the graft is completely replaced by the patient's own cells. Someday hospitals will keep frozen cultured tissue—grown from such sources as the skin removed during circumcisions—to use right away when severely burned patients are brought in.

Another option is synthetic skins, which have a fibrous dermal layer made of collagen, usually taken from cowhide or even shark cartilage. A sheet of silicone or plastic serves as the protective top layer that is eventually peeled off and replaced with grafted skin over the newly grown dermis.

Skin grafts—human, animal, or synthetic—decrease the amount of scar tissue that grows over the wound. While the human body has all these mechanisms for healing itself, it sometimes doesn't know when to quit. The deeper the wound and the longer it takes to heal, the worse the scarring will be. Scars can continue to grow and distort the wound for as long as two years. Age, skin color, genetics, infection, and hormones affect the growth of scars. Growing children, dark-skinned people, and pregnant women tend to have thicker scars.

It is a tribute to the adaptability of this important organ—some 18 square feet of surface area and fully 17 percent of body weight—that it can put up with all the daily abuse we give it and also fight back so courageously when it's injured. Scientists suspect that other human cells could be induced to grow in the laboratory and replace injured or diseased tissue in the same way skin cells do. This will take time, researchers say, and a better understanding of how different cells work.

CIRCADIAN RHYTHMS

Most living things follow a complex composite of daily, monthly, seasonal, and annual cycles. A human's daily cycle, when left alone and not stimulated by external cues, will continue for a period of roughly, but not precisely, 24 hours. The word *circadian* derives from that simple reality and two Latin words: *circa*, which means around, and *diem*, which means day.

All you have to do to appreciate the power of your own body rhythms is to start working the night shift or take a plane trip across several time zones. Fatigue, as well as eating and digestive disturbances, can result. Some studies indicate that disturbed body rhythms are also related somehow to depression, mental illness, and other problems.

Traveling across the time zones causes changes in heart rate and blood pressure, which can take a day to adjust. Body temperature may take as much as a week to return to normal, while certain hormones and enzymes may not return to normal for two or three weeks. The enzymes that are released to help the body digest food, for example, end up late for dinner.

There are some common misconceptions about the effects of tinkering with the body clock. For example, it's long been thought that shift workers who alternate between night and day work suffer from chronic fatigue and the accompanying damage to long-term health. Scientists have learned, however, that while some processes suffer when the body is asked to keep hours that are out of sync with its established pattern, others seem to benefit. Epidemiological studies in Sweden have shown that certain shift schedules actually lengthen lifespans, while studies done with mice have found that reversals in schedules tend to decrease the occurrence of mammary or breast tumors.

Chronobiologists, scientists who study the time cycles of living things, look for such connections between body time and health. Their research may result in better ways to predict and identify problems. Properly reading the internal rhythms of individuals also may lead to more effective treatments, because there seem to be times when medications and surgery are more successful.

One study using computerized monitors followed a group of children, recording everything they did, and was able to predict accurately which of them was depressed. These devices are now being used to study other brain malfunctions, including Alzheimer's disease, schizophrenia, and strokes.

It has long been thought that light plays a central role in coordinating body rhythms, and research studies have shown the connection. Seasonal Affective Disorder Syndrome, or SADS, is a kind of depression that occurs in winter when days are shorter and has been attributed to a lack of light exposure. By exposing volunteers to light/dark cycles of different lengths, researchers found that they could actually make adjustments in the subjects' body rhythms. Time intervals of extrabright light caused the subjects' body-temperature cycles to change. Body temperature goes down at night with metabolism and rises throughout the day until it peaks during midafternoon. Hormones follow distinct rhythms, too, some reaching their highest levels in the body during the night and some during the day.

Light is not the only important factor in the coordination of body rhythms, however. Experiments during the 1970s and 1980s, in which volunteers were squirreled away in isolation chambers for a month or more—far from sunlight, alarm clocks, and work schedules—showed that the body keeps its own schedule of sleep and activity that is only slightly different from the 24-hour period by which we usually manage our lives.

Most living things, including some single-cell organisms, seem to have these cycles, in which body temperature, metabolism, exercise, social habits, sleep, and other factors play a role. The interdependence of these factors complicates the search for the origin of circadian rhythms. And while light clearly plays an important coordinating role, the real direction appears to come from a complex network of glands and other tissues inside the body.

During the late 1980s, researchers found that two tiny areas at the base of the brain seemed to have a critical influence on human circadian rhythms. Suprachiasmatic nuclei (SCN), two dots located in the hypothalamus, have receptors for melatonin, a hormone that is linked to body time. Melatonin is released by the pineal gland, a light-sensitive organ in the brain, which studies have shown is important in the expression of circadian rhythms.

Researchers experimenting on hamsters discovered that by transplanting the SCN from one animal to another, they could alter the recipient's cycle to match that of the donor's. The removal of the SCN significantly altered the animal's body rhythms. Tumors located near the SCN in humans have profoundly affected their circadian rhythms. In another experiment, a British researcher found that melatonin, given to people who were traveling across time zones, could ease the problem of jet lag.

Other cycles besides the daily ones are influenced by melatonin. Numerous studies have shown it to be an important factor in animals' reproductive cycles, which can be delayed for months with doses of the hormone. Human metabolic cycles may change with the seasons as well, as evidenced by the tendency to gain weight in the winter and lose it in the summer. (Apparently, there is more to warm-weather weight loss than the desire to look good in a swimming suit.) Certain features of circadian rhythms may be inherited, and other experiments with hamsters have linked the length of an animal's cycles to a specific gene.

While investigators seem to be zeroing in on the organs that are important to the body's time structure, they have yet to uncover the activities inside the cell that are at the heart of it all. Experiments with cells taken from animals' pineal glands and SCNs show that the activities of these organs are directed by functions in the cells, and while the search for the secret to circadian rhythms continues to narrow, thanks in part to studies of fruit flies, sea snails, and other animals, the puzzle has yet to be solved.

MOLECULAR CLOCK

For many years, scientists calculated the evolution of humans and other creatures by examining the bones they left behind. The evolution of human intelligence was measured by the tools and other artifacts that were buried alongside these ancient hominids. Then in the early 1960s, American chemist Linus Pauling (1901–) and Austrian geneticist Emile Zuckerkandl (1922–) began a revolution in the study of the human family tree. They suggested that the story of human evolution could be read in the progress of molecular changes in the blood. These differences accumulate at a regular rate in each species, the researchers said, and could be used to measure time, just like any other clock, although over greater periods.

The molecular clock, as it's called, is based on the assumption that life on Earth had a common beginning, and that the evolution of species can be clocked according to the number of molecular changes that separate one from the other. Over time, mutations take place as species adapt to their environment and the challenges that life presents. As each takes a different path, mutations build up, creating different branches on the family tree of life.

Researchers set out to prove this notion by comparing proteins in the blood of apes with that of humans. Their aim was to find out how different they were and also figure out when and where on the family tree human evolution split from that of apes. Paleontologists had decided, based on the fossil evidence, that their paths had diverged some 30 million years ago.

Early experiments compared the molecular differences between humans and apes by injecting rabbits with blood from both. The rabbits formed antibodies to each: Antibodies are produced by animals, including humans, to match precisely the molecular structure of the invader, so that the next time it comes to call, specific antibodies are waiting to destroy it. In this case, the rabbit's antibodies produced to kill the human invader had a 96 percent response to chimpanzee blood. The researchers concluded that this 4 percent difference represented the divergence between human and chimpazee on the evolutionary tree. They went on to test many other creatures and came up with a family tree that closely resembled the one paleontologists had constructed from evidence in the fossil record.

These chemical comparisons revealed, however, that apes and humans reached a fork in the evolutionary road much later than paleontologists had thought, suggesting that human evolution branched away from that of apes as recently as 5 to 7 million years ago. Most paleontologists refused to take the idea seriously until fossil evidence was found to support it. In the meantime, progress in unraveling the chemical structure of DNA provided a powerful tool for dating the history of human evolution.

By unzipping the double helix in human DNA and gorilla DNA, for example, then stirring and heating it together, researchers were able to construct a double strand, one strand of which was human and the other gorilla, and see what matched up. They matched pretty well, as it turned out, and the differences were measurable. In this way, the researchers were able to plot more precisely the branches on the family tree and see that chimpanzees shared an evolutionary branch with early humans longer than gorillas, orangutans, or gibbons had.

Using another kind of DNA in the human cell, researchers have traced modern humans back to what they say is the one mother-of-us-all.

III

EARTH, THE DELICATE WEB

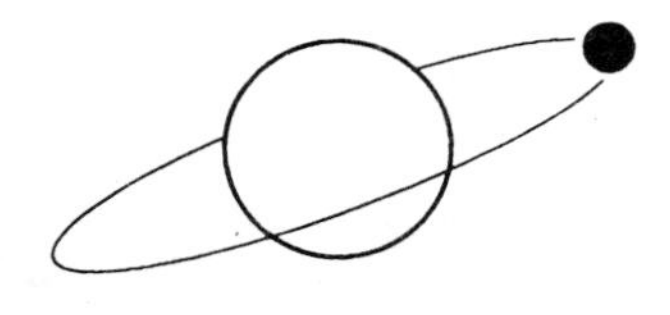

THOSE TRAVELLERS LUCKY ENOUGH to view Earth from afar can see that the shell of life embracing our rocky planet is surprisingly thin. Nevertheless, for over 3 billion years—despite the insults of asteroids, comets, and shocking climate change—life has managed to endure. Individual species may last only a few million years and ecosystems may ultimately fail to thrive, but the thread of life goes on.

It is increasingly apparent to scientists that life's tenacity, like that of a spider's web, relies on its ability to respond intricately to disturbances. Earth and the life on it seem to be an interconnected system in which a tweak at one spot causes distant threads to tremble.

Atmospheric gases, water, soil, rocks, and sunlight, along with a stunning variety of living things, make up the biosphere of Earth. The word *biosphere* is from Greek words meaning sphere of life.

When astronauts look down on this planet from space, they see it through a 600-mile-thick, cloud-streaked and luminous envelope of gases, which is a critical part of the biosphere. The atmosphere—which means sphere of vapor—serves many purposes. It is a storehouse of carbon dioxide and oxygen, a vast heat exchange system that regulates the temperature on Earth and drives the weather, and is a transfer medium for energy, water, and many living things, including humans. It also serves as a shield that keeps out the sun's harmful ultraviolet rays and admits only the energy that serves as the central power source for all life on Earth.

Sunlight is the chief energy currency in the biosphere. Solar energy is converted into chemical energy by green plants. This complex process, called photosynthesis, turns carbon dioxide and water into carbohydrates, locking some of the sun's energy in the bonds that hold the carbohydrate molecules together. Oxygen is released as a by-product of this process. The animals that eat the plants use oxygen to burn the carbohydrates, releasing the energy stored in the molecular bonds, along with carbon dioxide and water. In this way, carbon flows through the biosphere; if nothing goes wrong, the right amount is maintained in the atmosphere, while the rest keeps the cycle of life going.

The ocean serves as a holding tank or sink for dissolved carbon dioxide, helping to regulate the amount of carbon in the atmosphere. Much of the biosphere's carbon is locked up in biomass—the forests, plants, and other living things on the face of the Earth. Sometimes the energy housed in plants ends up in deep storage for hundreds of millions of years. Instead of being consumed by other living things or released when the plants decay, it's turned by time and pressure into coal and oil. The burning of these fossil fuels releases the warehoused solar energy as well as tons of carbons and other substances.

Whether plants are eaten, decay, or become fossil fuels, the energy they contain is transferred in a kind of bucket brigade in which some of the energy gets spilled. Every transformation diminishes the total amount of energy available for use. This situation is described by two important laws of nature.

The First Law of Thermodynamics states that energy can be neither created nor destroyed. In living our lives as part of the carbon cycle, we only borrow the sun's energy for a while and pass it on. However, as the Second Law of Thermodynamics states, the transfer of energy is not totally efficient; some of it gets lost to the atmosphere as heat that leaves the Earth as infrared radiation.

The energy balance sheet this represents must be carefully maintained in order for life to continue on Earth. Too much or too little energy received or shipped out could lead to dangerous imbalances. the Earth could become too cold or too hot to be habitable. Past imbalances in this energy exchange have played an important role in the extinction of many forms of life, including the dinosaurs. Current worries about global warming have to do with the possible consequences of human tampering with the Earth's energy accounts.

Along with the carbon cycle, of which photosynthesis is a part, there are other great cycles and ingredients that are necessary for the recipe of life. The hydrologic cycle, driven by solar energy, carries water from the surface of Earth into the atmosphere and back down again. Water molecules—released by plants and animals and evaporated from the water that covers 60% of Earth's surface—

rise to form clouds which are blown by the wind to other places, where they drop their moisture back to Earth.

Nitrogen, which makes up about 80% of our air and is an important part of the amino acids of which proteins are made, cycles through the biosphere with the help of microorganisms that convert its atmospheric form into one that is useable by plants.

Phosphorus also cycles through the biosphere. but unlike carbon and nitrogen, which use Earth's biomass, fossil fuels, oceans, and atmosphere as their reservoirs, phosphorus has its home in the rocks. It leaches slowly from stones and enters the food web. The mining of phosphorus and its use in inorganic fertilizers and detergents has vastly increased its activity in the biosphere and upsets the natural balance of aquatic food webs. It causes algae to bloom out of control. The excessive vegetation sinks to bottoms of lakes where increased bacterial activity robs the deeper, colder layers of oxygen. That, in turn, kills the cold-water fish, like trout and pike, and further upsets the delicate balance of the web. The lake ages and dies before its time.

The biosphere is divided into regions or communities that are called biomes—desert, tundra, grasslands, tropical rain forests, deciduous forests, and coniferous forests. Within each biome there exist networks of other communities called ecosystems. Each ecosystem is a complex of habitats for specific organisms that all play a role in a particular food web. The ebb and flow of energy through biomes, ecosystems and food webs causes shifts and changes that permit some organisms to thrive while others fade into history.

All of this wondrous complexity rides upon great stone rafts, the mobile plates into which the Earth's crust is divided.

PLATE TECTONICS

The outer shell of Earth is a thin crust of granite and basalt that rides on a sea of molten rock. The crust, or lithosphere (from the Greek word for stone), is divided into plates that carry the continents and oceans on their backs. The movement of these plates, including their interactions, is called plate tectonics (from the Greek word *tekton*, meaning builder or carpenter), and it determines Earth's changing face.

The story of how Earth sorted itself into layers of stone and metal begins with the formation of the solar system. When the sun was born from clouds of gas and dust, extra shreds of matter swirled in a ragtag parade on the hem of the giant star's skirt, and gravity whipped these accretions of matter into planets. Then the material that made up the planets settled itself into layers according to weight. The heavier elements in the Earth—nickel and iron—became the solid core of the planet with a radius of 2,600 miles. The core is surrounded by a layer of molten metals whose temperature causes the next layer, a mantle of rock, to flex and roll like a muscle. It is the boiling action of the liquid metal layer that is responsible for Earth's magnetic field.

Scientists think that the heat in the core is caused by radioactive decay of elements locked there. Some of it may also be leftover energy from pressure exerted when the planet formed. Because of the size of Earth, such energy hasn't dissipated as it has from our moon and smaller planets in the solar system.

The core and mantle layers account for 99 percent of Earth's mass. Riding on the mantle, like plates of armor, is the outer crust, a 60-mile-thick jigsaw puzzle of independent sections (seven big ones and a lot of little ones), which constantly jostle, scrape, and overlap each other. They move around on the surface about as fast as your fingernail grows. New plate is produced by cracks where molten rock rises from below and constantly pushes out new crust from either side of the crack. At other places, old plate is pulled back into Earth to melt again.

The movement of these plates sculpts the visible crust of our planet, drawing long scars and pushing up enormous wrinkles. Continents and oceans ride on the surface of the plates, having been shaped and reshaped over millions and millions of years by movement underneath. The changes are very slow and barely discernible to humans. We are most aware of the dynamics of Earth's structure when earthquakes occur and volcanoes erupt at the edges of plates.

Until the 1960s, many scientists thought the continents were an-

The Position of the Continents 250 Million Years Ago

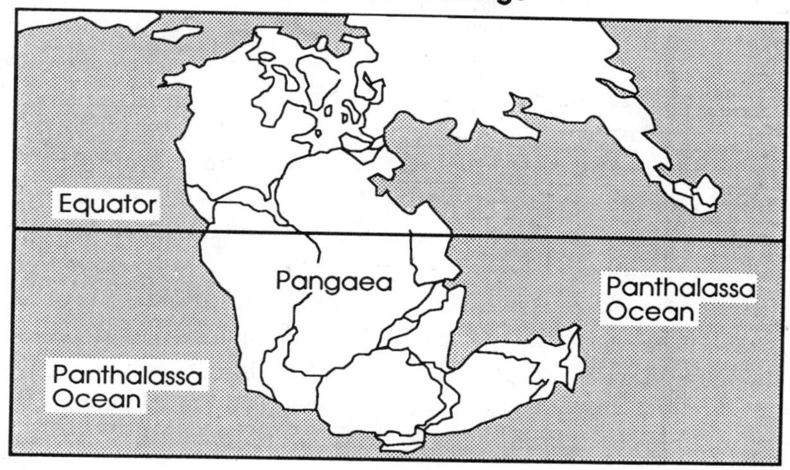

chored in place on intractable stone feet and that any movement in Earth's crust was up and down. A few thought otherwise, but no one paid much attention, even though, when explorers and mapmakers drew the outlines of the continents in the Western and Eastern Hemispheres, it became obvious that the two could fit neatly against each other. Benjamin Franklin suggested that the great land masses and oceans floated like rafts on some dense and powerful subterranean sea whose movement could easily buckle and break the rigid outer crust. He was very close to the truth.

It was the complementary shapes of the Eastern and Western Hemispheres that led Alfred Wegener (1880–1930), a German scientist, to propose a theory he called "continental drift." In 1912 Wegener suggested that the continents did, indeed, drift across the face of the earth to their current locations, after having broken away from one large landmass some 180 million years ago. He named the supercontinent Pangaea.

Pangaea itself was thought to have been formed long before, when large landmasses had crashed together. As time went by, the conglomeration of land gathered smaller chunks to itself until it became a supercontinent. Until recently, scientists thought it had stayed that way, a C-shaped landmass that straddled the equator for more than a hundred million years and then broke apart into the shapes we know today.

Wegener's idea was ignored at first. As time went by, however, it began to make sense to the scientific community, for it was based on evidence that could not be explained in any other way. The same fossils of prehistoric reptiles were found in such far-flung locations as India, southern Africa, and Antarctica. Certain trees lived on widely separated continents, as did flightless birds with similar characteristics. Scientists had earlier proposed the existence of land bridges to explain these things.

In the 1960s, several researchers, working in different ways, built on Wegener's theory and all came to the conclusion that the solid Earth, as we like to call it, is an ever-changing thing. Europe and North America are moving away from each other an inch or more per year; India is bashing head first into Asia, pushing Tibet out of the way, just as Indochina was shunted to the side long ago by the same collision. This same confrontation also produced the Himalayas.

Geologists have now refined the original idea of the footloose landmasses. It's thought that, over the millenia, chunks of real estate have broken apart, drifted around, and bashed into each other on a regular basis, and that today's landmasses consist of many blocks of crust that were once free agents. For example, geologists have found long stripes of crust in North America that once belonged with land on the other side of the Atlantic Ocean. But the ways in which continents were formed and where the puzzle pieces orginated are still hotly debated.

The power behind the jostling plates is convection currents, in

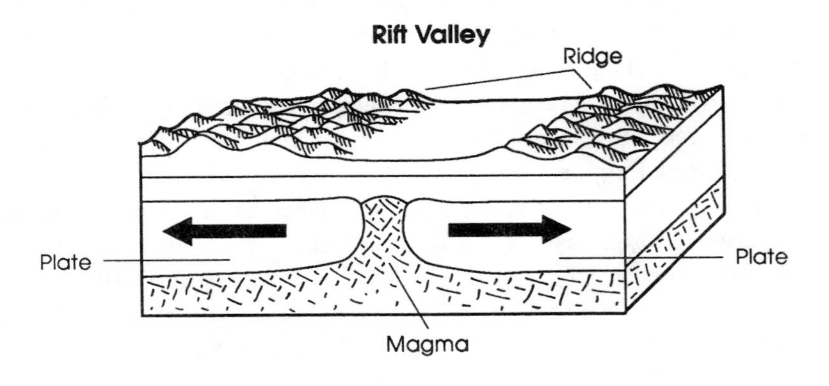

which heat is transferred globally through the mobile layer of hot rocks in the earth's mantle, or asthenosphere, a word derived from a Greek term meaning weak. The hot rock deep in the mantle tends to rise, and the cooler, denser rock near the outer crust tends to sink. The consequences of this boiling effect are felt where the plates meet.

These mobile sections of the earth's crust are also pushed against or under each other by the force of new plate surface emerging from midocean ridges where molten rock rises from a giant crack in the planet's skin. The crack, called a rift valley, rolls out new ocean floor regularly. There are also places where the boundaries of divergent plates occur underneath landmasses. For example, the Great Rift Valley in Africa is a crack where two plates are pulling apart. Another rift lies under Siberia, where it created the oldest and deepest freshwater lake in the world, Lake Baikal.

The discovery that midocean ridges preserve a record of Earth's magnetic field has provided scientists with some convincing evidence for the existence of plate tectonics. Just as an audio tape records a magnetic signal by changing the direction of the magnetism of metallic flecks, the new ocean floor makes a recording of Earth's magnetic field as it cools. Oceanographers knew that the magnetic field experienced reversals over the millenia, and they discovered that the orientation of the metal particles in the basalt found along rift valleys corresponded to changes in the magnetic field. Maps of these magnetic stripes look like zebra skin.

It became clear that midocean ridges were the point at which plates diverged and where new crust was being created. These plate boundaries snake through the middle of the Atlantic, around the horn of Africa, up through the Indian Ocean to the Arabian Sea, down around Australia, and north through the Pacific. The ridges and rift valleys are interrupted by faults or cracks that allow blocks of plate to grind past each other, creating sharp angles. Under parts of the ridge there seem to be chambers of molten rock that have oozed up from the deep. Scientists think these storerooms of magma supply the crust-manufacturing process, as well as the volcanoes that dot plate boundaries.

Sometimes one plate slides under another and becomes a part of the molten underworld again. This process, called subduction (from

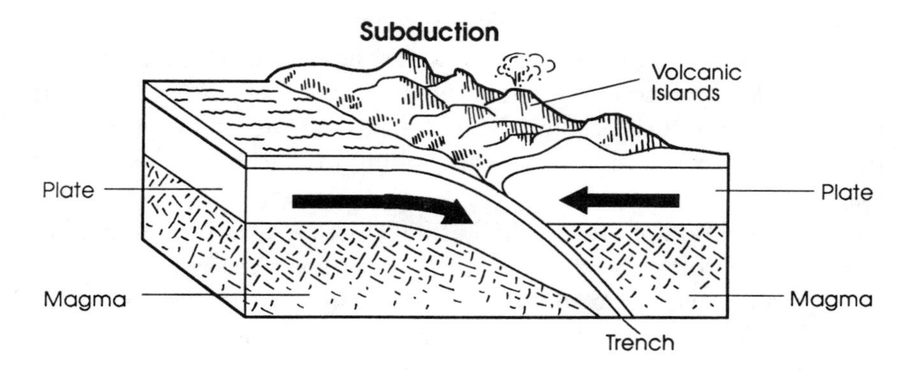

a Latin word that means "to lead under"), can take millions of years and occur in several ways. When an ocean-bearing plate crashes into one that carries a continent, the ocean crust dives down into Earth, while a deep trench forms along the edge of the more resistant landmass. Such is the case along the edges of Chile and Peru. When the plate and mantle resist each other, cracks form, causing deep earthquakes. Melted rock or magma spurts through the surface of the earth as volcanoes.

Subduction zones can also occur in the middle of the ocean, creating green, underwater mud mountains. As diving plates melt, they release water, which mixes with minerals in the mantle and oozes up through cracks in the ocean floor as a mud volcano. These greenish mile-high mountains seem to have the consistency of cream cheese.

Plumes of hot rock also burst through cracks near subduction zones under the ocean and create huge mountain peaks that are dragged away by the moving plates and transformed into island chains. The Hawaiian Islands were formed in this way. Volcanic islands can also form in an arc along the point of collision between ocean and continental plates. This is the case along the western edge of the Pacific plate, where island arcs include Japan, Indonesia, New Guinea, and New Zealand. In fact, because they are studded with volcanoes and endure 80 percent of the earthquakes in the world, the edges of the huge Pacific plate are referred to as the Ring of Fire. Some of the world's biggest cities lie near the seismically active Ring of Fire.

Earthquakes also occur at the edges of plates that are just trying to slide past each other. The plates become blocked and build up pressure, causing the rocks to bend and fold. The strain opens cracks called faults, which crisscross plate boundaries, and sudden releases of pressure between the sliding blocks of rock cause the ground to heave and roll. Devastating earthquakes have happened along California's coast, where the Pacific and North American plates meet, as well as on plate boundaries and faults in Mexico, Afghanistan, Iran, Turkey, China, and Armenia.

Since it is impossible to travel down through the layers of the earth and investigate them, it is fair to ask how scientists know all this about the structure and behavior of the planet. Seismology—the study of sound waves as they travel through the earth—has been the primary tool. Seismologists are able to determine the shape and composition of the earth's layers by examining the different shapes of soundwaves as they travel outward from earthquakes or explosions. The various densities of Earth's layers reflect and refract the sound waves differently. Vibrations in the ground are collected by a seismograph and converted into a graphic representation, which

The Ring of Fire

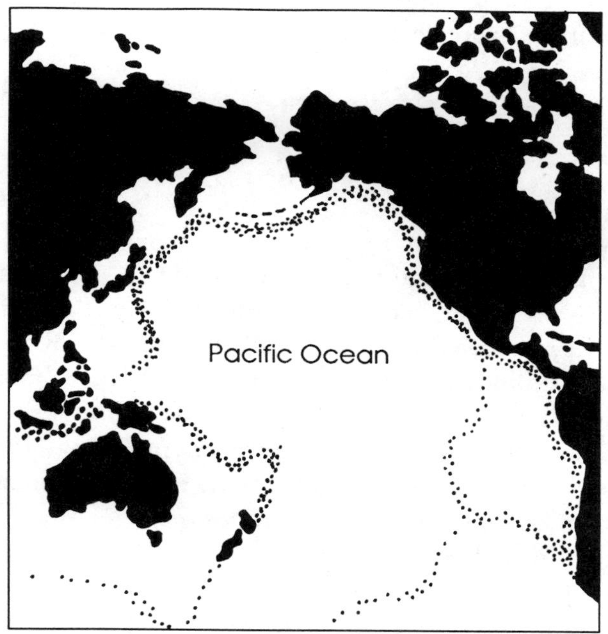

seismologists are trained to interpret. Seismographers, with "listening" stations spread out over the globe, are also able to keep tabs on the underground testing of nuclear weapons. Computers have also been essential to a better understanding of the architecture of inner earth, making sense of all the data scientists collect and organizing it into three-dimensional images or maps.

Advances in other technologies have been important in the development of the plate tectonics theory as well. The discovery of mid-ocean ridges and their role in shaping Earth's surface would not have been possible without the development of submersibles that could go deep into the ocean and collect information. Sophisticated sonar systems that operate from ships on the surface have been useful in the task of mapping the ocean floor; about 5 percent has been mapped so far. And the ability to take deep-sea core samples of basalt along the ridges has supported the notion that plate material was being created there. Oceanographers have found that basalt farther away from the ridges is older than that close by.

A renewed interest in the nineteenth-century study of the geometry of spheres, as well as the ability of satellite technology to "see" the big picture of a changing Earth, have also contributed to the current theory of plate tectonics.

RICHTER SCALE

When you hear the Richter scale mentioned in news stories every time there is an earthquake of any consequence, the number used to describe it is deceptively low. That's because the Richter scale—invented by California seismologist Charles Richter (1900–85)—is logarithmic, which means that every additional number on the scale multiplies the effect by 10. The lowest tremble that 1930s equipment could measure was assigned a 0 value; an earthquake with a magnitude of 5 would be 100,000 times stronger than 0. Modern seismometers can detect the slightest trembles in the earth, with these baby quakes assigned negative numbers on the Richter scale.

The world's most devastating earthquakes have approached 9 on the Richter scale, causing terrible destruction when the ground moved in waves. In 1985, an 8.1-magnitude earthquake hit southern Mexico, killing nearly 10,000 people, injuring 20,000, and damaging or destroying buildings in a 300,000-square-mile area. The tremors were felt in Texas.

The magnitude of an earthquake is a meaningful measurement of the destructive power of such an event, but it is also a relative matter. The structure of underlying rock and soil affects the severity of the shock, as does the way in which the buildings are made. The extreme damage suffered by portions of Mexico City has been attributed to the layer of clay that lies under the center of the city: It trapped and amplified the waves produced by the quake.

The earthquake that struck Armenia in 1988 registered 6.9 on the Richter scale with a 5.8 aftershock, and yet it was one of the most destructive in recent years, destroying whole towns and killing thousands of people. An earthquake measuring 7 on the Richter scale struck San Jose in 1989 but claimed far fewer lives. Most of the sixty or so people who died were trapped by a collapsed elevated highway in San Francisco 60 miles away. The big quake, like the one that hit San Francisco in 1906, is yet to come.

HYDROTHERMAL VENTS

The deepest floor of the ocean is like a desert, or so scientists have always thought. Most living things depend upon the energy of the sun to sustain the food chain, which begins with photosynthetic plants. But the sun cannot penetrate much deeper than 600 feet into the ocean, so photosynthesis is out of the question. It was thought, therefore, that not much could be alive in the deepest deep.

That notion changed in 1977 when a 16-ton submersible vessel called *Alvin* went down to explore a midocean ridge near the Galápagos Islands in the Pacific Ocean. Scientists aboard the vessel were amazed to see rich communities of strange undersea life flourishing in what should have been a dark, sterile underworld. They found crabs, mussels, clams, and colorful fish living near clusters of giant tube worms with flowing red tendrils. What they learned was that these communities are supported by underwater hot springs, spouting from vents in the Galápagos Ridge a mile and a half below the surface of the ocean. Hydrothermal vents, as they are called, are created when ocean water flows in and out of cracks in the rocky seam of the earth and is warmed by heat from the mantle.

Unlike most life on earth, the creatures living near hydrothermal vents do not depend on photosynthesis. They are part of a food web whose foundation is sulfur; it is consumed by a kind of bacteria that uses it and other chemicals in the water to manufacture proteins, in a process called chemosynthesis.

Perhaps the strangest of all the life-forms discovered near hydrothermal vents is the giant tube worm that has no mouth or digestive system. The five-foot-long creature is nourished by a huge internal factory that is composed of billions of bacteria. The worm absorbs raw materials—hydrogen sulfide, oxygen, and carbon dioxide—and the bacteria do the rest, producing food to nourish the worm. When researchers examined the samples scooped up and carried to the surface of the ocean, they discovered that the worms had red blood, rich in hemoglobin, as did the clams, which spat red blood at their captors. The sulfurous foundation of these creatures' food supply also gave them an unmistakably rotten smell.

There are hydrothermal vent communities at other ridges in the Pacific and Atlantic oceans. In 1990 scientists found the first-known freshwater hydrothermal vents in Siberia's Lake Baikal, the world's oldest, deepest lake, which was formed by a rift under the Asian continent and contains 20 percent of the world's fresh water. It's not yet clear whether the ecosystems surviving near the Lake Baikal vents are based on chemosynthesis.

Southwestern Oregon's deep-blue Crater Lake is also the site of hydrothermal vents, although researchers did not find giant tube worms and other strange creatures that have been observed in vent communities elsewhere. However, large mats of bacteria growing in the unusually warm water at the bottom of the lake indicate that chemosynthesis is taking place there. A high salt content in Crater Lake has led scientists to believe that minerals from the earth's mantle have been leaking from cracks in the crust, causing the unusual clarity of the lake. They also found old, inactive mineral "chimneys" that indicate where active vents had existed in the past.

Strange vent creatures have also been discovered in places where there are no vents at all. These spots, called cold seeps, occur where petroleum and other substances leak out of the earth and provide the chemicals needed for chemosynthesis.

Scientists have wondered how all the creatures that inhabit vent and cold-seep communities were able to spread through the world's deep waters to these widely separated locations. A recent discovery off the coast of California may have provided the answer. Scientists aboard *Alvin* found the same kinds of chemosynthetic ecosystem living off the carcass of a huge whale. The current speculation is that the hundreds of whales and other marine animals that die every year provide niches for bacteria and other creatures; they also create stepping stones across the ocean floor that let bottom-living animals move across the deep oceans.

Hydrothermal vents may also be sites of the world's richest mineral deposits. The hot water circulating through the rocks at the ocean bottom dissolves metals like iron, copper, silver, and tin and blows them out into the cold water of the ocean. Drifts of particles fall to the ocean floor in deposits estimated to contain millions of tons of valuable metals, awaiting the technology that can mine them.

BIOLOGICAL EVOLUTION

When Charles Darwin (1809–82) published *On the Origin of Species* in 1859, he started an argument that rages in some quarters to this day. Whereas scientists debate the question of how evolution occurred, some conservative Protestants insist, despite overwhelming scientific evidence, that the universe and everything in it was created "as is" from nothing, as recently as 10,000 years ago.

Darwin started the fuss in 1858 by proclaiming—along with another scientist, Alfred Russel Wallace (1823–1913)—that life on earth has gradually changed as organisms have adapted to their environment. Darwin described evolution as the struggle for existence that occurs when species produce more offspring than the environment can support. Variations within species allow some individuals to adapt more successfully to their environment and survive long enough to produce offspring. Successful characteristics are promulgated within the species, allowing those individuals that possess them to thrive, while others disappear. Over time, these adaptations accumulate until the species has evolved into something quite different from its origin, and the changes are visible in the fossil record.

The idea that life on earth has evolved is not original to Darwin. Renaissance scholars suspected this was true and worked on the problem that thinkers as early as Aristotle (384–322 B.C.) had considered. René Descartes (1596–1650), a French philosopher and mathematician, wrote about evolution and then retracted his statements, perhaps to avoid being accused of heresy. Philosophers, geologists, and other scholars in the seventeenth, eighteenth, and nineteenth centuries, including Darwin's grandfather, wrote and lectured on the possibility that life had evolved on earth. But it was Charles Darwin who captured the steady unrolling of life in a set of principles that profoundly affected the way scientist look at the world.

Darwin, however, was not able to explain the *mechanism* by which species passed on their successful characteristics. A few years after the publication of Darwin's ground-breaking book, Austrian monk Gregor Mendel (1822–84), who was not formally trained in science, explained this mystery. By quietly crossbreeding peas in a small garden in the monastery, Mendel worked out a set of laws that form the foundation of the science of genetics. He found that characteristics are passed on as units of inheritance that do not blend with or change each other. They appear in successive generations according to mathematical laws; some are dominant and appear, while others are recessive and show only if both genes, as he called them, are recessive.

The chemical nature of genes was not understood until two Cambridge University biochemists—James Watson and Frances Crick—deciphered the chemical structure of DNA, the molecule that carries the message of inheritance.

PUNCTUATED EQUILIBRIA

Darwin's widely accepted notion of gradual change—gradualism—has been challenged on all fronts by researchers, most recently and controversially by Stephen Jay Gould and Niles Eldridge, two American paleontologists who have proposed that evolution occurs in bursts of change, punctuating long periods of equilibrium.

Paleontologists have always been frustrated with the gaps in the fossil record. The bones and imprints of creatures with obvious connections to the past seem to leave no clear evolutionary trail. How did they get from here to there? Darwin himself expressed frustration with the lack of transitional evidence, and decided that the links had been there but their remains were not.

Gould and Eldridge, on the other hand, suggest that the gaps exist because the links were never there. Instead, new species arose from isolated groups living in the outposts of their ancestral line's territory. Separated from the gene pool of the larger group, they evolved quickly along a different path, and emerged so completely different that they can't reproduce with the original species.

The fossil record in the old neighborhood appears to contain a gap from which the transitional stages are lost. What actually happened, according to Gould and Eldridge, is that the members of the new species reentered their ancestral territory and, not being able to mix with their long-lost relatives, lived and died there as a new species.

Gould and Eldridge also suggest that species should be regarded as individuals that are born and die, as individual creatures do. The individuals within a species are small parts of a greater whole, and might be looked upon as trees in a forest or blood cells in a great red stream. Just as skins, shells, and exoskeletons are the boundaries that make living things appear to be individual, so do climate, ocean currents, and geography form boundaries for species.

So evolution seems to occur on several levels. The day-to-day adaptations to environmental circumstances cause the characteristics of individuals within a species to oscillate between relatively narrow boundaries; trends are represented by accumulations of these adaptations over many years. Every once in a while, special circumstances allow the status quo to be interrupted by the relatively sudden appearance of something new. Finally, the slow ticking of the evolutionary clock is abruptly interrupted by periods of great dying. Mass extinctions—at least eight during the past 600 million years—serve to reset the evolutionary clock by wiping out more than half of the species alive at the time.

MASS EXTINCTIONS

Extinction is the greater part of life on planet Earth. Nearly all of the things that have lived here are gone. We who are left, the 40 million species alive today, represent less than 1 percent of all the species that have occupied the planet. Every year a hundred or more species disappear, a pattern that has continued steadily since the emergence of life on earth. This ongoing background extinction occurs for any number of reasons, but primarily because of changing climate, the loss of food and habitat, or the invasion of some new predator.

Rising above the steady dying-out of species, there are short periods, relatively speaking (up to several million years long), during which many more species disappear. These mass extinctions are characterized by the disappearance of entire groups of widely varying species from different kinds of habitats. The survivors are also affected by whatever cataclysmic event caused the suffering; their numbers are greatly reduced.

The fossil evidence for mass extinctions has provided geologists with a framework for the ages of the earth. The great geologic ages and periods are bracketed by what appear to be overwhelming periods of disaster. The nearly trackless causes of these events are obscured by time and erosion, but scientists have devoted their careers to uncovering them.

They have outlined a series of mass extinctions, the first of which occurred some 440 million years ago at the end of the Ordovician Period, wiping out approximately one-fourth of the species living. Eighty million years later, during the Late Devonian Period, it happened again. The largest mass extinction, called the Great Dying, occurred 250 million years ago at the end of the Permian Period. It is estimated to have wiped out almost 96 percent of the species on Earth at the time—a close call.

Nevertheless, it is interesting to note, some scientists say, that over the ages, the ebb and flow of life has kept some kind of balance sheet, so that extinctions have not outpaced the emergence of new life. In fact, they have served to weed the garden so that new forms could evolve. And despite what must have been horrific events, life has not been extinguished.

Some researchers have suggested that Permian extinction happened when Earth crossed paths with a flurry of killer comets, knocked from their distant home outside the solar system by a wandering star. Deadly climactic changes, and the accompanying loss of food and habitat that seem to have occurred during the Great Dying, have also been attributed to the explosion of a nearby star. There isn't much evidence for either explanation, however, and the most likely explanation was change in climate and sea level.

During the Late Triassic Period, about 202 million years ago, 20 percent of the existing species were eliminated, and then 66 million years ago, some 60 percent of the living things, including the last of the big dinosaurs, were lost. While there were other less devastating extinctions (as many as twenty major ones reflected in the fossil record), it is the curtain call for the "terrible lizard" that is now most intriguing to the public.

DINOSAURS

Dinosaurs—ranging in size from a small chicken to half the length of a football field—lived on this earth for 140 million years. Researchers are learning that these diverse and mysterious creatures cared for their young, ran like elephants, and sniffed out their prey like the cleverest fox.

And a few of them actually survived. On any good day you can see them, if you look into the sky or search the branches of trees. Birds are now thought to be the modern descendants of carnivorous dinosaurs that lived millions of years ago, and scientists are discovering that dinosaurs had many of the same features that birds have, including feathers, which are modified scales. The oldest bird in the fossil record, Archaeopteryx, was a one-foot-long carnivorous amalgam of dinosaur and bird that lived 145 million years ago in what is now Europe. For years some paleontologists argued that instead of being a bird, this creature was just a dinosaur with feathers, but Archaeopteryx is now widely accepted as the first bird.

At a recent conference in Germany, most of the attending paleontologists agreed that birds are clearly derived from dinosaurs. Scientists had been arguing over the comparison since the 1920s when a researcher first suggested the relationship, based on observed similarities. Pterosaurs, however, with their long beaks and leathery wings, weren't birds but airborne reptiles. During the 1970s and 1980s, renewed interest in the bird-dinosaur problem led to discoveries with the help of modern technology, as well as good old-fashioned digging.

Birds were thought to have evolved from a specific branch of small, sleek dinosaurs, far removed from the giant meat eaters. But the CAT scan, a kind of X-ray that provides three-dimensional images of body interiors, has revealed interestingly birdlike structures inside the skulls of even the biggest meat-eating dinosaurs, such as the fearsome Tyrannosaurus Rex, five tons of bad news that certainly couldn't fly.

T. Rex and its relatives appear to have had birdlike air chambers and nerve pathways inside their heads, too.

The recent discovery of hundreds of dinosaur nests, eggs, and nestlings—so rare that scientists didn't expect to find enough of them to draw any conclusions—has strengthened the connection. More than five hundred eggs and a herd of thousands of dinosaurs were uncovered in the badlands of Montana, providing evidence that duckbill dinosaurs nested together and cared for their young until they were big enough to fend for themselves and avoid being trampled by their parents' big feet.

The nests, similar to those of trumpeter swans and sandhill cranes, were big mounds of mud with scooped-out hollows to hold the eggs. They were built a dinosaur length away from each other, suggesting that these animals might have lived as part of a community and a family. These findings contradict the image of the cold-blooded lizard that lays its eggs and walks away.

Image has been a problem for the dinosaur, whose name derives from Greek words meaning "terrible lizard." The most important new development in the study of dinosaurs may be the emerging change in attitude towards these misunderstood creatures. Recent finds, as well as a new look at museum specimens, are changing the image of the dinosaur from a plodding evolutionary failure to a complex, dynamic, and socially active creature.

CAT scans reveal that carnivorous dinosaurs may have had acute senses of smell and hearing, making them clever hunters. Their skulls were engineered to be lightweight and efficient killing machines. And, like birds, they may have hooted at each other with calls that some researchers compare to woodwinds or horns. Computer analyses of the size and shape of vocal cavities in dinosaur skulls have led them to compare dinosaur herds to brass bands.

Researchers have taken a fresh look at the way dinosaurs were put together, and how they stood, walked, and ran. They've examined the tracks that were left in the mud and preserved by sediments for millions of years. This new information is eroding the image of the lumbering, clumsy, lizard reduced to spending its life buoyed by water. All the evidence points to even the biggest dinosaurs as being capable of chugging along like angry elephants, some with their tails held high.

That's the latest thinking about Diplodocus, a graceful, giant herbivore that lived at the beginning of the Cretaceous Period, about 144 million years ago. The first nearly perfect specimen was uncovered in the late 1800s and reassembled for display in Pittsburgh. But it wasn't until recently, at the Science Museum of Minnesota, that the animal was mounted with its tail held aloft, an arrangement that paleontologists now agree is the right one, based on the shapes of the bones.

It's not surprising that in the study of dinosaurs—animals that have been dead for at least 65 million years—old ideas are only occasionally overthrown. There's very little evidence left on which to build a hypothesis. It takes a lot of conjecture and creative thinking to piece together the bare bones of creatures we hardly know.

Most of the diverse population of dinosaurs were not preserved for scientists to examine. Like most of the other living things that have occupied the earth, dinosaurs' bodies decomposed or were eaten by other creatures. Only thin slices of time were trapped in sediments that turned to rock. Scientists have cataloged as many as 1500 species so far, as small as chickens and as large as the 140-foot-long Seismosaurus, or "earth shaker," discovered recently in New Mexico. Even so, that may be less than half of the number that really existed.

Strange new discoveries are being made every day, though, especially in Australia, South America, and China. Some seem to have had weird anatomies, such as the pug-faced and oddly built carnosaurs found in South America, and Mongolia's toothless, parrotlike creatures with unusually long forelimbs and short tails.

Some questions about dinosaurs may never be answered scientifically. Certain physiological characteristics—color, markings, and plate arrangements—are probably lost forever. But the positioning of things like tails and feet can be inferred from the shapes of the bones. A new look at the bone structure of Diplodocus, for example, has convinced many paleontologists that this creature had been misunderstood. Now they believe that this long-necked herbivore stood on its toes like an elephant and used its tail, fully half its body length, not only as a powerful whip but also as an instrument for shedding surplus heat, just as humans sweat and dogs pant to cool down. The energy it would take to carry such a tail aloft and brandish its 15-foot whiplash might have required a high-pressure metabolism, suggesting that the dinosaur was a warm-blooded animal, another recent and controversial idea that may never be settled.

Some dinosaurs may have been both cold- and warm-blooded, like modern leatherback sea turtles that are known to stoke some mysterious internal fire to raise their core temperature when surrounded by icy water. Such an arrangement may have been necessary for the animal's survival.

But something happened 66 million years ago that the big dinosaurs couldn't survive. The most dramatic explanation is a six-mile-wide meteor hitting Earth. Scientists have found evidence of two impacts, one in the Gulf of Mexico and the other in Manson, Iowa. The Gulf impact came first, in mid-June, scientists think, setting off immense tidal waves and blowing material sky high, blocking out the sun and causing freezing temperatures. Two to four weeks later, a second impact left a 30-mile-wide crater in Iowa and scattered quartz sand grains all over North America. The impacts may have come from one huge object that split apart and made two hits and scientists from several disciplines have studied the blobs of melted glass, fossils of strangely damaged plants, dust layers, sea sediments, and craters that are clues to this one- or two-stage cataclysmic event.

Nevertheless, the impact theory itself remains controversial, since a giant meteor wasn't the only blow to fell the dinosaurs. Many of them were already gone by the time the meteor hit, done in by changes in climate and sea levels, food shortages, and small mammals that ate their food. After the meteor hit, the oxygen content of the air was reduced by 10 percent, a serious problem for big animals.

The mass extinction that wiped out the dinosaurs (and more than 60 percent of the other species on the planet) has acquired new importance lately. The prospect of another chunk of space junk hitting Earth has stimulated some scientists to suggest unusual defense plans. For example, rockets carrying nuclear bombs could race out to meet a threatening asteroid, explode nearby, and deflect it from Earth's path.

In addition to this renewed interest in the way dinosaurs died, there is also an active interest in how they lived and what sorts of plants and animals shared Earth with them. Paleontologists are busy painting in the background of the dinosaurs' world, which was governed by the same natural laws that rule ours. The increasing importance of the dinosaurs' environment also influences the way in which fossils are exhibited in museums. Exhibitors are interested in presenting these creatures as biological organisms that were products of the same evolutionary stream that produced us and every other form of life on Earth.

By any evolutionary scorecard, dinosaurs were a great success, dominating the planet for 140 million years, much longer than humans have existed. Indeed, most species average about a million years. It was the demise of the big dinosaurs that made more room for mammals like us to evolve. Together we are part of a vast stream of life that stretches across millions of years into today. After all, dinosaurs still roam the earth, represented by almost nine thousand species of birds.

TROPICAL RAIN FORESTS

Tropical rain forests are the richest and most important ecosystem on the planet. They grow near the equator in warm, moist areas of Africa, South and Central America, and Southeast Asia. Trees as tall as 150 feet create a green canopy through which dappled sunlight falls onto the vines and small plants, which usually cling benignly to them and to the smaller trees and shrubs in their shadows. The foundation of this tangled green mass is the network of roots, fungi, and thin acidic soil of the forest floor. Nearly all of a tropical rain forest's nutrients are locked up in vegetation, making it an efficient biotic engine that lives off its own debris, not off the well-drained soil in which it's anchored.

Tropical rain forests grow primarily at low altitudes. However, some cling to the sides of mountains as high as 4,500 feet. Their trees are shorter, hung with moss and perpetually shrouded in clouds. Hence, their name—cloud forests.

This ancient web of life, the tropical rain forest, has been called Earth's lungs because it absorbs huge quantities of carbon dioxide and exhales oxygen back into our atmosphere. It's also a sophisticated water pump, slowly releasing heavy rainfall (eight or more inches per month) into rivers and streams, while cleansing and recycling over half of the rain back into the atmosphere. One-fifth of the planet's fresh water is recycled annually by the Amazonian rain forests alone. All this moving water creates a persistent cloud cover that reflects solar energy, regulating Earth's temperature and influencing global weather patterns.

Tropical rain forests also preserve the ancient genetic blueprints for most of the living things on our planet. Many of our most important crop plants, now a result of human as well as nature's engineering, have their beginnings on file in ancient forms still living in these forests. Botanists often have to go back to the rain forest to get germ plasm, the reproductive cells involved in heredity, from these native plants in order to revive exhausted or infested crops.

It is no small concern that every year, millions of acres of rain forest are being cleared for lumber, flooded by hydroelectric dams, or burned to make way for agriculture. By shrinking the rain forests, these practices may contribute to the greenhouse effect in which carbon dioxide traps solar radiation inside the atmosphere and raises the average temperature of the planet (see GREENHOUSE EFFECT, page 156). Rain forests also hold captive billions of tons of carbon that, when burned, are released into the atmosphere to cause mischief as carbon dioxide.

Alarmed by the rate at which these forests are shrinking, a group of tropical biologists has formed a team that will conduct speedy surveys of the remaining forests, ranking them according to their biological diversity and deciding which are the most threatened by human activity. This Rapid Assessment Program (RAP) has been somewhat controversial among biologists, but the members of the RAP team contend that the loss of world rain forests has accelerated so quickly that there isn't time for the more lengthy, traditional study. They view forests from airplanes, use satellite imagery, and explore on the ground to track the fates of these ecosystems, and have already found hundreds of different species of plants, animals, and insects living in specific forests and nowhere else.

Over half of all the animals in the world live in tropical rain forests, which hold more diversity per acre than any other ecosystem. One small area may hold thousands of different insects, animals, trees, and other plants. These forests are also the winter homes for many of the world's birds, which fly south for the winter, and so the deforestation of that habitat may very well mean the extinction of many species, which in turn alters biological evolution forever. Because of its minute complexity, this fragile ecosystem, once damaged, takes as long as two hundred years to regenerate.

Over millions of years, all tropical plants, birds, and mammals have formed symbiotic relationships. Early attempts to bring some plants out of the forest failed because early explorers didn't recognize the production partnerships these plants had formed. For example, the orchid that produces the vanilla bean requires a specific bee to pollinate it.

Because the dense vegetation and hot, humid conditions of the rain forest have resisted human exploration, many species living there are totally unknown. It's been estimated that five of six tropical species have never been seen by scientists or probably anyone else. Important new food crops and vital pharmaceutical products may be hidden in the rain forests, too. Only one of ten known tropical plants have been examined so far for their medicinal qualities. Even now, 25 percent of our prescription drugs come from the tropical rain forest, revolutionizing some areas of medicine. For example, the rosy periwinkle, found on Madagascar, has improved the chances of children with lymphocytic leukemia; and heart surgery was made possible by curare, a muscle relaxant that South American Indians use as arrow poison.

Until recently, the forest canopy, which houses 80 percent of the rain forest's inhabitants, was largely unexplored. Scientists got at its riches by cutting down the trees or blasting shot guns into the air and examining what fell out. They have now built floating rafts and strung catwalks along the roof of these beautiful green oceans. And they are giving names to exotic creatures we have never seen before, while finding medicines, food crops, and beautiful specimens to take back to the rest of the world.

OZONE

Ozone has an unfortunate double personality. It is a highly reactive ring of oxygen atoms that causes great mischief close to home and yet, when far away in the stratosphere, serves as a protector.

Most of the ozone that pollutes our cities comes from a chemical process that is set in motion mainly by the burning of fossil fuels. Hydrocarbons and nitrogen oxides from auto emissions and smokestacks react with water vapor and ultraviolet radiation from the sun to create smog. One of the dangerous chemicals in smog is ozone, a volatile greenhouse gas that also damages crops, trees, and other vegetation by attacking cell membranes. In people it makes breathing difficult and compromises the immune system. The Clean Air

Act of 1970 established limits for air pollution, but efforts to reduce the hydrocarbons emitted by cars and factories have had little effect on the ozone levels in many American cities.

Some, like Los Angeles, are doomed to have problems because their geography and weather conditions combine to trap the smog.

Other cities may suffer because they have too many trees. Vegetation sheds hydrocarbons, mainly highly reactive olefins, by the ton every day, reaching a peak during the hottest hours. Research has shown that olefins are efficient smog producers. Cutting down all the vegetation is not the solution, of course. What's needed, some researchers say, is a reduction in the levels of nitrogen oxides, one of the other ingredients in the poisonous ozone soup pot.

Up in the stratosphere, though, ozone is a hero. It forms a thin layer 15 miles above the surface of the earth that serves as a vital shield between the sun and us, blocking out 99 percent of the ultraviolet radiation, the portion of sunlight that damages DNA. In fact, life on earth probably did not begin until this protective filter developed a billion years ago.

Ozone is created in the stratosphere when sunlight breaks up oxygen molecules into separate oxygen atoms, which then are strung together with other oxygen molecules like beads in fragile bracelets. Ultraviolet radiation from the sun breaks apart the bracelets, releasing spare parts that are restrung into new bracelets. Fortunately, this constant recycling of oxygen and ozone molecules ties up a lot of ultraviolet radiation, preventing it from hitting the earth.

The materials for this stratospheric manufacturing process, however, are in limited supply—about three parts per million. Any reduction in the supply upsets the delicate balance of ozone production and reduces the molecule's ability to block ultraviolet radiation. Unfortunately, something is destroying the ozone. Scientists have been measuring ozone levels in Antarctica since 1956, and satellites began monitoring stratospheric ozone in 1969. During the following decade scientists became increasingly alarmed at the depletion rate their tests were showing. They claim that chlorofluorocarbons (CFC) are to blame.

CFCs are odorless, human-made gases that are used in the manufacture of computer chips, insulation, packaging materials, and other foam products, as well as in air-conditioning, industrial solvents, and propellants. In the 1970s, efforts in the United States, Canada, Norway, and Sweden to reduce the use of CFCs as propellants for aerosol sprays were successful but did not reduce the destruction of ozone. And despite worldwide efforts to curtail production of these chemicals, the depletion of atmospheric ozone continues at a rate of 4 to 5 percent per year. In 1985 a hole as big as the continental United States was discovered in the ozone layer over Antarctica; scientists believe it began forming in the late 1970s.

The air over Antarctica is particularly suited for the breakdown of ozone and the creation of giant holes in the shield it forms. CFCs float easily into the atmosphere from our homes and businesses and are whirled into the Antarctic by global air streams. There they are trapped by a giant vortex of wind rising from the cold ocean currents circling the continent. Ice clouds form within the walls of this trap at the beginning of the Antarctic winter. The clouds' ice crystals act as catalysts for a chemical reaction between sunshine and the CFCs, releasing chlorine. When spring arrives the chlorine eats a hole in the ozone layer. Some researchers think that sulfates from volcanic eruptions and burning fossil fuels contribute to the problem year-round by eliminating nitrogen compounds, which control the chlorine.

Interestingly, the hole over Antarctica has been worse during odd-numbered years when winds at the equator reverse direction and blow from the west. Nevertheless, each year the hole gets deeper, and there is evidence that smaller holes are developing over the Arctic, as well. Overall, global levels of ozone continue to decrease. By 1991, winter depletion of the ozone over the Northern Hemisphere had grown to 5.6 percent, and observers recorded summertime depletions from 2.9 to 3.3 percent.

In 1991, the northern winds that normally sweep ozone into the hole were late, and the opening in the protective shield remained well into summer. That year Antarctica was blasted by twice the usual level of solar radiation. As a result, scientists have observed a 12 percent decline in phytoplankton, the single-cell floating plants that are an important source of nutrients for other sea life and that form the foundation of the marine food chain. By changing the climate and interfering with plant life at the cellular level, ultraviolet radiation could reduce land crops as well, and world food supplies would suffer.

In humans, the increase of ultraviolet radiation reaching earth could also increase the incidence of cataracts, with 1.6 million new cases by the turn of the century. Skin cancers could double in the next forty years, with 300,000 new cases worldwide by the year 2000. And while sunscreen lotions and darker skin pigmentation may protect some people from skin cancer, they won't prevent suppression of the immune system and the infectious diseases resulting from overexposure to ultraviolet radiation.

Faced with all this frightening information, the global community has agreed to stop making or using CFCs by the year 2000. Some CFC producers, in light of recent evidence, have pledged to stop production even earlier. But CFCs seem to have a long life, and even if there were no more of them produced or used on the planet, the chemicals that have leaked into the air for years will continue to do damage. The CFCs released today will take a decade to reach the stratosphere and then remain there, destroying ozone for almost a hundred years. Some researchers worry that it is already too late to save the ozone layer.

Ideas for intervening in what seems like a desperate situation include somehow sending human-made ozone up to the stratosphere. Unfortunately, the energy costs for doing so are staggering. One scientist has even suggested using huge infrared lasers to zap airborne chemicals before they can reach the stratosphere. Meanwhile, manufacturers have developed CFC substitutes that are supposed to break down before they reach the ozone layer. Although they contain less of the chemicals that make CFCs dangerous, they still are capable of destroying ozone. And ironically, these substitutes are hydrocarbons that may contribute to pollution problems on the surface of the planet.

Oddly enough, new evidence suggests that, for a time, city dwellers may be protected from the ravages of ultraviolet radiation by pollution, while the loss of ozone in the lower stratosphere may actually counteract the warming effect of ozone loss at higher levels. This and other twists in the ozone problem make it a critical and puzzling challenge for world survival.

GREENHOUSE EFFECT

If it weren't for the greenhouse effect, our earth would be a very cold place, with an average temperature of about 0°F instead of the more livable 63° that it is now. But scientists are worried that our planet may be suffering from too much of a good thing. In recent years, human activity has contributed record amounts of gases to the atmosphere, creating a chemical shell that holds in heat and threatens to change the world's climate.

Just as sunlight passes through the glass roof of a greenhouse, solar radiation, in the form of short wavelengths, passes through the atmosphere that surrounds the planet. Fortunately, ozone in the stratosphere turns away most of the dangerous ultraviolet radiation that can wreak havoc on plants, animals, and humans. However, like the contents of a greenhouse, the warmed earth radiates heat in the form of longer infrared wavelengths. Water vapor and certain other substances in the atmosphere absorb these longer wavelengths, trapping heat, rather like the way in which the glass in a greenhouse keeps the wind from stealing the heat trapped inside. Carbon dioxide, nitrous oxide, methane, and ozone are the substances called greenhouse gases.

Natural processes on earth have produced greenhouse gases for millions of years, as part of a complex system that maintains life. For example, carbon dioxide acts as a transport vehicle in the carbon cycle, trucking this essential element from biological warehouse to factory to warehouse in an endless loop. Carbon dioxide is a byproduct of respiration in animals and bacteria, and is used by plants to make sugars for energy. It is constantly cycled through life forms, with some of it drifting into the atmosphere and some going into various kinds of storage facilities.

All the living things on earth together store about 600 billion tons of carbon. As this living matter dies and decays or is burned, carbon dioxide is released back into the atmosphere. Almost twice as much carbon lies trapped in dead material, some of it captured and stored by sea animals that carry it to the bottom of the sea when they die. Earth's largest reservoir of carbon is contained in fossils of ancient organic matter that got stuck in the mud, so to speak, where bacteria could not break it down to release its CO_2 into the atmosphere. Over millions of years this material changed into coal, oil, and oil shale. When these fossil fuels are burned, staggering amounts of CO_2, about 5 billion tons per year, are added to the atmosphere. Destruction of forests adds another 2 billion or so tons of CO_2 yearly.

Nitrous oxide (N_2O), ordinarily produced by bacteria, figures into another complex cycle that circulates nitrogen through Earth's systems. Its growing presence in the atmosphere is due to combustion of plant and fossil fuel and the use of nitrogen fertilizers in agriculture.

Methane, an odorless, invisible gas, is produced by matter that decays in the absence of oxygen. It is the main component of the natural gas we use to heat our homes and is produced naturally by bogs and wetlands, as well as in the digestive processes of cattle, sheep, and termites. Humans may have doubled the amount of methane in the atmosphere as a result of landfills, rice paddies, mining, natural gas exploration which leaks methane into the air, and the raising of cattle and sheep. Ordinarily methane, along with CO_2 and carbon monoxide, is scrubbed from the atmosphere by a chemical housekeeper called hydroxyl. Apparently there is now so much of these gases that hydroxyl can't keep up.

Adding to the naturally occurring greenhouse chemicals are carbon monoxide, produced by the combustion of fossil fuels, and chlorofluorocarbons, human-made chemicals that are used as solvents and propellants, as well as in refrigeration and air-conditioning.

The danger of a greenhouse gas varies according to its ability to trap infrared radiation and its staying power in the atmosphere. Some of these chemicals remain in the atmosphere for only a few years. Methane, for example, has a life of about ten to twelve years if hydroxyl is available to clean it from the air. Chlorofluorocarbons, on the other hand, can remain in the atmosphere for as long as a hundred years. Scientists fear that future generations will have to pay for our use of these chemicals.

Since the beginning of the industrial revolution 150 years ago, human activity has drastically increased the concentration of greenhouse gases in the atmosphere and interfered with the natural systems that keep them in balance with each other. According to environmental scientists, human tinkering with this chemical balance threatens to make equally drastic changes in the earth's climate.

GLOBAL WARMING

During the 1980s, the earth experienced six of the hottest seven years of the century, seemingly part of an overall global increase of 1°F in the last hundred years. That doesn't sound like much, but many scientists fear that the human contribution of greenhouse gases to the atmosphere is tipping the balance that the earth keeps among its chemical and thermal accounts. The penalty, they say, could be harsh.

As the earth warms up, the ocean acts as a buffer, soaking up heat and keeping the planet cool. It can't absorb all of the increase, however, and the current trend, when plotted into the future, may mean an increase of another two to five degrees over the next hundred years. That still seems inconsequential until you realize that a decrease of nine degrees in the average temperature of the earth 18,000 years ago caused glaciers to crawl over a big part of the Northern Hemisphere. A warming of less than that magnitude could change our habitats in a number of very important ways.

Higher global temperatures would cause the water in the oceans to expand. The polar ice caps would also begin to melt. As a result, beaches would disappear and coastal wetlands, which serve as breeding grounds for sea life, habitats for birds, and buffers against storms, would be destroyed by the salt water flooding into them. And, at a time when they might be most necessary, coastal wetlands around the world have been destroyed to make way for development. Harbors and ports would be ruined and coastal cities eventually destroyed by higher sea levels, and the ferocity of tropical storms, which serve as heat transfer systems for the global climate, would increase, causing even more suffering and financial loss.

Significant global warming would change rainfall patterns, with some arid and semiarid regions receiving more precipitation, and some food-producing areas receiving less than they need. The life cycles of tree crops that depend on cold periods would be altered. Insects that usually are killed off in winter would flourish. Increases in carbon dioxide could alter plant chemistry in such a way that insects would be forced to eat more leaves to get the same amount of nutrients. Farmers might be tempted to use more petroleum-based insecticides, but that, scientists warn, would only contribute to the greenhouse effect that caused the problem in the first place.

All of these changes could have an important impact on plant, animal, and marine species. Some will be able to adapt; many others won't. Coral reefs, for example, are very sensitive to changes in the temperature of the sea water they inhabit. Their destruction would mean a loss of habitat for species of marine life and the loss of protective barriers against the pounding of the sea.

The burning of coal and other fossil fuels—the major warehouses of carbon on earth—seems to be the biggest contributor of carbon dioxide to the atmosphere and the chief culprit in global warming. Atmospheric levels of CO_2, measured at the Mauna Loa Observatory in Hawaii since 1957, have risen by about 20 percent during that time.

The second major source of increased carbon dioxide in the atmosphere is the burning of forests and other vegetation, contributing to the problem in more than one way. Vegetation of all kinds clears carbon from the atmosphere and uses it in photosynthesis, and so when we burn forests, we lose that potential for clearing the air, so to speak, and add a lot more carbon while we're at it.

The link between greenhouse gases and global warming was further illustrated by research in Antarctica. Scientists took samples from the layers of compacted snow and ice that have been building up there for thousands of years. Rather like examining the growth rings of a tree, they studied the core samples taken from the ice to find out more about conditions on the planet at various times. The amount of CO_2, methane, and other substances trapped in ancient air bubbles deep in the ice corresponded with the periods of cold and warmth in the life of the planet, confirming what climatologists believed, that the amount of CO_2 in the atmosphere seems to correspond directly with global temperature. This research also shows that concentrations of CO_2 are the highest they have been in the past 160,000 years.

The earth does have coping mechanisms that can counteract much of the damage that humans cause. The oceans and the atmosphere cooperate in giant feedback systems that work to equalize imbalances. However, scientists fear that if we tip things too far off base, some previously unobserved and fearsome feedback process could be unleashed. The long-term changes by which the earth tries to resume equilibrium may not be healthful for many of the species living here, including humans.

Some scientists believe that a CO_2-rich environment will enhance the growth of many plants, especially trees, which will in turn remove more CO_2 from the atmosphere. Studies have not been able to prove this hypothesis, however. On the contrary, global warming may thaw northern soils and tundras, setting free the Co_2 they've kept in the deep freeze and out of circulation.

Atmospheric scientists are also investigating the link between global warming and changes in the energy output of the sun. For years, researchers have tried to show a correlation, but they are limited by the lack of data. Variations in solar output have been accurately plotted for less than two decades. The best climate-change data cover only a small portion of the globe, the Northern Hemisphere, and have been collected for less than 150 years. However, more recent studies have found the closest-ever correlation between global temperatures and changes in sunspot activity. Scientists won't know for sure until they have collected data on changes in the solar constant and global temperatures for years to come.

The truth is, we are far from understanding the complex interdependencies of ecosystems. We can't be sure what will happen if global warming continues. And although some researchers do not agree that the earth is in trouble from greenhouse gases, most are sufficiently worried to encourage immediate action. The long life of certain substances in the atmosphere may mean that the warming process can't be stopped soon in any event. But by taking immediate action to reduce greenhouse gases now, environmental scientists suggest, ecosystems and humans will have more time to adjust to the inevitable changes. They are urging the nations of the world to find other energy sources besides fossil fuels, and at the same time work to preserve and replenish the forests.

ACID RAIN

The thought of acid raining out of the sky is pretty frightening. But even though you are not in danger of being burned by acid if you walk in the rain, the conditions that cause it can be dangerous if you have such respiratory problems as asthma, bronchitis, and emphysema. Acid rain is also dangerous indirectly because it increases the acidity of our environment, upsetting the pH balance.

Rain is normally slightly acidic, because carbon dioxide in the air dissolves into rain water, giving it a pH of 5.6. The pH is measured on a logarithmic scale, which ranges from 0 to 14 with 7 being neutral. (On a logarithmic scale, each number jumping up or down multiplies the effect by 10.) Acid rain is anything below 5.6 on the pH scale.

Rain is only one of the forms in which acid-laden moisture falls to earth and does mischief. Acid falls in the form of snow, hail, sleet, fog, and dust, too. Acid deposition, as it's called, is caused by the combustion of fossil fuels, which produces sulfur dioxide and nitric oxides that, in turn, become sulfuric and nitric acids.

Sulfur is a brittle, yellow substance that used to be known as "brimstone," as in "fire and brimstone." It's a part of most proteins (like fingernails and hair), and is present in varying degrees in fossil fuels. When oil or coal is burned, sulfur is released into the air mainly as sulfur dioxide (SO_2), a smelly and irritating gas that on its own damages lungs, as well as other living tissue like trees and plants. But when SO_2 combines with ozone, nitrogen oxides, and hydrocarbons in the presence of sunlight, it is oxidized (that is, it acquires another oxygen atom) and becomes sulfer trioxide (SO_3), which in turn becomes sulfuric acid when exposed to water. If there is no rain or other precipitation, the sulfur trioxide can drift many miles downwind and fall as acid rain somewhere else.

Nitrogen oxides are produced by the chemical interaction of nitrogen and oxygen during the combustion of fossil fuels, mainly in furnaces and cars. This process sets in motion a complicated cycle of reactions between sunlight, oxygen atoms, and hydrocarbons, another byproduct of combustion. The end result is the accumulation in the air of ozone and nitrogen dioxide (which gives smog some of its characteristic hazy-brown color). Both are irritating to eyes and lungs and destructive of plants. Nitrogen dioxides, in the presence of water, turn into nitric acid and fall as acid rain.

Sulfur and nitrogen oxides from the American West and Midwest are carried eastward by the prevailing winds and fall as acid rain on the northeastern United States and Canada. Thousands of Canada's lakes are so acidic that fish are unable to breed and live in them; Canadians blame 75 percent of their acid rain on the United States. The same problems exist in Europe. Norway and Sweden claim that more than half of their lakes are dead because of acid rain. Great Britain reports similar damage to its lakes. Acid rain also seems to be killing trees in forests around the world.

Acid deposition directly affects whatever it falls on, but acid compounds also soak into the soil where they can dissolve substances such as lead, mercury, aluminum, and cadmium. These metals can then leach into lakes, rivers, and streams, threatening the things that live there, as well as the humans who get their water supplies from these sources. Acid precipitation seems to damage not only the leaves of trees and other vegetation but also their root systems, where soil acidity affects the way plants get nutrients. Their leaves turn yellow and fall; eventually the plant or tree dies.

We know that acid deposition is a worrisome danger for ecosystems, but it has also melted the noses off of stone statues and is a serious threat to art, architecture, and historical artifacts all over the world.

RADON

Radon is an odorless, tasteless, and invisible gas that can get you where you live. It leaks from rocks and soil into your home by way of foundation cracks, drains, and other openings. In lesser amounts is it also found in building materials, well water, and natural gas. The Environmental Protection Agency (EPA) attributes 20,000 lung-cancer deaths per year to radon gas exposure. But it is not actually radon that causes the problem. Radon's daughters are to blame, and you'll find out who they are in a moment.

Radon 222 is the product of a chain of radioactive decay that begins with uranium 238, an unstable element that exists to some degree in most of the rocks and soil on the earth. It is unstable because it does not have enough force to hold together all of its nuclear particles. So, something has to go. In the case of uranium 238, it's two protons and two neutrons. When uranium 238 or any other radioactive element decays, it means that something happens in the nucleus of the atom to spit out particles, changing the atom to something else—another element, or daughter. (A daughter, in this sense, means the product of the radioactive decay of an element.)

The unwanted protons and neutrons cling together in what is called an alpha particle through a process called alpha decay. The atom that was uranium 238 becomes thorium 234, the process continuing inside rocks for several generations of elements. Some of the elements experience beta decay, in which two particles are pushed out—a neutrino and a kind of electron. They are the product of a strange change inside the nucleus, in which a proton becomes a neutron, or vice versa. Some elements in the decay chain emit gamma rays, highly excited waves of energy. Radon has a half-life of almost four days. A half-life is the time it takes half a given number of atoms to decay. The half-lives of the other elements of the chain vary from mere seconds to billions of years. Uranium 238 has a half-life of 4 billion years.

Most of the decay that leads to radon, is harmless, locked up inside the rocks. At some point, though, radon 222 is set free, and being a gas, it's sneaky. It can leak into the spaces between rocks and grains of sand and soil. From there it can climb up into houses, which act as traps for the gas that would otherwise disperse harmlessly into the air. Every square inch of soil emits six atoms of radon every second.

When radon sneaks into a home, the occupants inhale it. Fortunately, it is exhaled long before it decays. The chance of one radon atom decaying in your lungs is pretty slim. But polonium 218, radon's daughter element and a solid particle, is fond of hitching rides back into the lungs on dust particles that float in the air. When inhaled, the particles carrying polonium 218 stay in the lungs for as long as half an hour. Polonium 218 has a half-life of 187 seconds;

it has plenty of time to decay and emit an alpha particle while it rests on the delicate tissues of the lungs. Polonium 214, another daughter that can be inhaled, is a few more generations down the chain; it decays in 200 microseconds, also shooting out a destructive alpha particle.

Smoking apparently adds to the problem by filling the air with fine particles that collect the polonium and keep it suspended in the air for many more hours than it normally would. This increases the risk of radon-related cancer to smokers and nonsmokers alike. Children may be especially vulnerable to the effects of radon because they inhale more, often increasing their overall exposure to the particles emitted by radon's daughters.

Although alpha particles have a very short range (two or three inches in the air or a fraction of an inch in human flesh), they are capable of cutting through the strands of DNA, the chemical blueprints, inside a cell. Beta and gamma rays also damage DNA, but not as messily as the clumsier alpha particle. Enzymes, a kind of chemical repair crew in the cell, are able to repair damage to the blueprints if only part of the double strand is broken; by using the remaining strand as a pattern, they are able to reconstruct the missing parts. If both strands are damaged, however, the crew is forced to patch things up as best they can. Any resulting changes or mutations in the genetic code can be passed along to new generations as the cell divides and replicates itself. This mutation, scientists believe, lays the groundwork for cancer. Later, because of smoking or some other insult, the cell or cells turn into cancer.

The amount of radon trapped in houses can be at least ten times higher than outdoors. Depending on the location of the house, it can be thousands of times higher. Certain parts of the country have rocks and soil types that emit more radon and allow it to travel about more easily. Granite, which contains a lot of uranium, underlies large areas of the United States. Phosphate and shale deposits also presage higher radon levels in homes. Of course, deposits of uranium drastically increase the risk of radon.

Sandy, loose soils allow radon to escape more readily, whereas clay inhibits its travel. Radon is also slowed down by water content in the soil, but it is capable of traveling for many miles by hitching a ride with groundwater. (This tendency is useful to seismologists, who are able to predict earthquakes by monitoring radon levels in groundwater, which fluctuate a few days before an earthquake and then rise sharply just before it begins.) Water from deep wells is more likely to be contaminated with radon than municipal water supplies, where water is stored long enough for radioactive elements to decay away.

The decay of radon *222* is measured in picocuries per liter (pCi/l), measuring the number of radioactive decays in a liter of air. A curie is 37 billion radioactive decays. A picocurie is 1 trillionth of a curie. One pCi/l of radon means 133 decays in a liter of air in one hour. The EPA has established 4 pCi/l (532 decays) as the level beyond which further testing should be done. A radon level between 20 and 200 pCi/l demands repairs at some point, while anything above 200 pCi/l should be attended to immediately. Solutions include sealing cracks and openings, installing ventilation systems that will carry the radon away from the house's foundation, sealing contaminated building materials, and installing charcoal filters to clean radon from well water.

Hundreds of companies sell radon detectors, which the homeowner can use to detect radon levels in the house. The EPA's National Radon Measurement Proficiency Program tests and rates the efficiency and reliability of these devices. The EPA recommends testing for radon in fall and winter when levels are higher because windows in a house are closed, trapping in the radon.

Some scientists say that the EPA has overestimated the risk from radon: Studies conducted by researchers at the University of Pittsburgh found that U.S. counties with the highest measured radon emissions had the lowest lung-cancer rates. Such findings raise questions about the actual risk of low-level doses of radiation. Many studies verify the danger of large radiation doses; calculation of low-dose risk is based on that information, with the risk proportional to the amount of exposure. Some scientists question a method that was based on studies of uranium miners and think that low levels of exposure may be much less dangerous than originally thought.

But if EPA estimates are correct, three out of a hundred people exposed to a lifetime level of 4 pCi/l of radon will die from cancer, a risk roughly equivalent to smoking half a pack of cigarettes or driving a car every day.

EL NIÑO

Ocean and wind currents course over the planet in distinct patterns created by the spinning of Earth and the heat exchange between the atmosphere and surface waters. Coming together at the equator, these currents become the trade winds that blow westward over the Pacific, pushing surface water toward Australia and Indonesia. The winds pull rain out of a moist low-pressure system over the western Pacific, giving India and Southeast Asia their monsoons.

As the warm water moves west, cold, nutrient-filled water from deep in the ocean rises near the coast of Peru, creating a rich fishing ground. Meanwhile, a high-pressure system sits over the southeastern Pacific near Tahiti. It works with the low-pressure system in the west to keep the trade winds blowing, making rain for Asia's rice and giving Peruvians plenty of good fishing.

In this global engine that determines climate, the phenomenon called El Niño appears to act like a switch. When it is turned on every three to eight years or so, weather disasters like drought, floods, and storms can affect parts of the world that are thousands of miles from each other. Peruvians call it El Niño—Spanish for "Christ Child"—because of an annual warming of the waters off the coast of Peru around Christmas time.

During the 1982–83 El Niño, the most destructive in this century, India, Southeast Asia, and Australia endured droughts, dust storms, and fires, because the wind had carried away the moist air that would have produced rain. Florida and the Gulf of Mexico's coast were drenched, and California suffered mudslides. Weather was affected as far away as Alaska and Africa. That particular El Niño indirectly killed thousands of people and cost billions of dollars.

An El Niño begins when the pressure systems over the Pacific reverse. Trade winds begin to fail and blow gusts back east, pushing warm water towards south America. The water warms the air above it, which rises and sucks more westerly winds into the area, keeping the system alive. This warm, moist air slams into the South American coast, bringing heavy rain. Fishing declines because the warm water prevents the flow of nutrients from the deep. The warm water damages coral reefs and upsets the ocean food chain.

Then, after El Niño has shuffled and redealt the world's weather, scientists now suspect, it lays the groundwork for the next one. The warm water that piled up near South America spreads north and south for thousands of miles, feeding into the system of currents that will eventually carry it back to the equator, where the whole thing starts over again in a few years.

Scientists are studying this giant feedback effect not only to understand how it works but also to be able to predict the occurrence and severity of El Niños so that people can prepare for them.

DESERTIFICATION

Every year more than 50 million acres of the earth's land are pushed past the point of supporting profitable crops. Fully a third is turned into desert, and the land that is most vulnerable is the already arid region that shelters delicate and easily trampled ecosystems.

We are most aware of this problem in the Sahel, a broad swath of land in Africa where devastating famines have killed hundreds of thousands of people. The African climate has experienced cycles of wet and dry conditions for the past 2000 years, but where areas are devastated by drought, the problem is magnified by the presence of human activity. In addition, huge areas of Asia, Russia, and the American West are turning into desert, too.

Deserts are not devoid of life, of course, but they cannot grow crops and provide graze for cattle and other farm animals. They are forced to do so by humans, more often than not, who ask too much of a fragile ecosystem. Human history is filled with tales of rich islands of fertility that were eventually overtaken by barrenness.

This process—called desertification—occurs most easily in arid and semiarid land when humans overuse or mishandle the resources in the area. Overgrazing by cattle, which has had devastating effects in Africa and the American West, destroys the grass and other plants in an area that already gets infrequent rains. Vegetation is stripped of its leaves and can't support the root systems that anchor the soil and absorb water. The animals also trample the earth around watering holes and streams, killing life there and compacting the soil.

When the rain falls it can't soak into the soil to feed the plants and recharge the groundwater. Instead, it runs off into streams and rivers and ultimately into the sea, taking productive soil with it. Sometimes, to compensate for losses incurred because of decreasing amounts of graze, cattle herders will increase their herds, which only accelerates the decline.

One important reason that cattle cause environmental change is that they occupy grazing land in unnaturally large numbers and aren't able to move around the way natural grazers do. The fauna that existed naturally on these lands, both in Africa and the American West, were a more natural mix of grazers and browsers. Browsers, like deer and antelope, simply have a different strategy for getting their food. They eat the leaves of bushes and trees, rather than concentrating mainly on grasses as cattle do. And because the number of cattle is kept artificially high to keep up with human demands, they have a more extreme impact on the landscape than a natural, more self-regulating system does.

With these conditions in place, a drought comes along and pushes the threatened ecosystem over the edge into disaster. The cattle die because of lack of food, and with the water shortages and crop failures caused by drought, humans also suffer and die in great numbers. Such ecological changes deprive the world of many species of plants and animals, which simply disappear.

With the loss of topsoil through wind and water erosion, the ecosystem is changed for a long time. It takes hundreds or thousands of years to build topsoil, and so the change from grasslands to arid ground that merely supports shrubs is nearly irreversible, from the human standpoint.

The areas that are most sensitive to overgrazing are the boundaries between two ecosystems. Environmentalists say such overgrazing by cattle in the past century has transformed the wide grasslands of the American West into unyielding desert. The 3 percent of America's beef that comes from the area, they say, does not compensate for the desertification of what is mainly public lands.

Improper irrigation is another factor in desertification. If irrigated land is not properly drained, it can become waterlogged and incapable of ridding itself of salts, which build up in the soil over time until nothing will grow there. The process is reversible up to a point, but environmentalists say farmers and governments usually prefer to abandon the land rather than work to undo the damage.

To make matters worse, drastic changes in ecosystems drive changes in climate, which perpetuate the problem. By stripping away plant life, desertification robs land of its vital connection with the hydrologic cycle, which processes water from the sky through the yielding and absorbent earth, and slowly back to the sky to fall again as rain. A desert, on the other hand, is a hot plate that cooks the rain out of the sky and keeps it from falling on nearby lands. And so the desert grows.

SUSTAINABLE AGRICULTURE

Soil—a thin and delicate crust of sand, clay, rotted vegetation, microorganisms, and other critters—is all we have to feed the world. Over thousands of years farmers have devised ways to nourish soil so that it will continue to nourish us. They've learned to till into the soil the animal and crop wastes that are converted by microorganisms into nutrients and stirred by worms and other creatures to create tilth, the ability to soak up and retain the rain, a characteristic of good soil. Through trial and error farmers learned to rotate crops; they confused the pests and also maintained the nutrient level in the soil by alternating nitrogen-producing plants, like legumes, with those that are mainly nitrogen consumers, like corn and wheat.

Plants can't use the nitrogen (N_2) that makes up 80 percent of the air. Somehow they have to "fix" the nitrogen, that is, convert it into ammonia (NH_3), a molecule they can use. This job is performed by bacteria that live in nodules or bumps on the roots of legumes such as alfalfa, soybeans, and peas. These bacteria combine nitrogen and water to produce the ammonia and oxygen that they need for growth. Luckily, legumes "fix" more nitrogen than they can use, leaving it in the soil to nourish the next crop to be planted there.

The development after World War II of chemical fertilizers, pesticides, and herbicides brought about a fundamental change in farming, however. The amount of land that one farmer could work was increased. This change in the "eyes to acres" ratio, the number of acres a farmer watches over, worked to reduce his or her ability to pay attention to the soil. It may have made farmers' lives easier and allowed them to produce more food, but it also changed the way they cared for the soil. Critics of what is now called conventional agriculture refer to chemical and mechanical power as an interference in the partnership between the farmer and nature.

Agrichemicals and fuel-intensive machinery also made farmers more dependent than ever on outside resources. Chemicals allow farmers to plant their acres repeatedly with the same crops. Such monocropping, instead of the well-planned rotation of mutually beneficial crops, invites disease and pests, requiring an ongoing increase in chemical use. And chemicals usually take the place of organic fertilizers, which add texture and stability to the soil; without that anchoring effect, soil is vulnerable to wind and water erosion. As a result, billions of tons of topsoil end up in the rivers, lakes, and streams every year.

Critics contend that although these changes produced record yields and allowed farmers to manage more acres, they also exacted high costs from the economy of rural America and from the environment. Meanwhile, the impact of chemicals on the ground water and changes in the quality of the soil have led some researchers and farmers to reconsider the benefits of conventional practices. Environmental scientists warn that the millions of tons of nitrogen fertilizer that are added to the soil each year are polluting lakes, streams, and rivers and also escaping into the atmosphere in the form of nitrous oxide, a greenhouse gas. (See GREENHOUSE EFFECT, page 156, and GLOBAL WARMING, page 158.)

One response to these problems is sustainable agriculture, which calls for a reduction in chemical use, the preservation and enhancement of the soil, and protection of the environment, along with the production of good food and profits. It is not a return to pre–World War II farming methods, but a marriage of old and new.

The benefits of modern equipment, in addition to the lessons of soil, seed, and livestock research, are combined with conservative practices that make the most of resources right on the farm. For example, animal and green manure—a high-nutrient crop such as alfalfa that is plowed under in order to decompose, providing texture and nutrients—are considered the mainstay of healthy soil, anchoring it against wind and water, while adding nutrients to feed the plants that sprout from its surface. When green manure and other crops are rotated according to their ability to consume or contribute nitrogen, sustainable agriculture maintains the soil by means of natural systems. Proponents say crop rotation also improves yields.

Diversification also includes intercropping, or planting several different crops in the same field, which allows plants with different needs to support each other by repelling each other's pests and weeds and stabilizing the soil. This is very different from the large fields of single crops that are characteristic of conventional farming.

Diversification is seen as insurance against the unpredictability of weather and markets as well. Rather than relying on the success of one crop, farmers can divide resources among several crops and livestock, thereby spreading the risk. Conventional agriculture has been concentrated on a few profitable crops, mainly wheat, corn, and soybeans; the world, however, is rich with thousands of species of food-bearing plants, which farmers are beginning to investigate. This sort of diversification could be insurance against the catastrophic crop failures that would result from climate change and disease.

Advocates of sustainable agriculture suggest that farmers can control pests and disease by manipulating their environment and making a home in the fields for natural predators and parasites. Ladybug beetles are a well-known example of a beneficial insect. When pesticides are used, they are targeted carefully to minimize the damage to such beneficial creatures and to the environment.

A 1980 U.S. Department of Agriculture report found that organic farming methods, which play an important role in sustainable systems, are productive and efficient with capital and resources. The report has opened the door for more research on alternative farming methods. Some of the findings show that farms practicing sustainable agriculture have crop yields competitive with conventional farms. Though yields are sometimes lower, the difference tends to be offset by lower production costs.

Critics of federal crop support programs contend that they discourage farmers from diversifying their crops; at the same time, USDA support for research on biological pest control has declined. It could be that farmers are understandably wary of anything that appears to be a return to the harder work and lower profits of pre–World War II farming. For whatever reason, conventional practices are still used by most American farmers today.

WETLANDS

Until recently, wetlands were considered a waste of space. During the previous century, they were even thought to be an unhealthy influence. Now ecologists fear for the health of the planet and those of us who must live here, because the stabilizing influence wetlands play in global ecology has been compromised by the efforts of agriculture and industry to claim these rich and diverse ecosystems for development and other purposes. Despite efforts to slow their loss, wetlands are disappearing by up to 400,000 acres per year.

The distinguishing characteristic of most wetlands, of course, is water—either as a shallow cover, punctuated by deep pools, or in saturated soil interwoven with living and decaying plants. They are often transitional areas between water environments and dry land, and are populated by many trees and plants that can put up with the occasional flood or, at the very least, don't mind having wet feet. They provide habitats for hundreds of different kinds of waterfowl, fish, amphibians, and mammals, as well as great populations of butterflies. Wetlands support some 45 percent of all endangered animals and 26 percent of endangered plants, according to the National Wildlife Federation.

The northern peatlands, or bogs, covering many millions of acres in North America, Canada, and Siberia, are wetlands that develop in cold climates, in basins or on flat or gently sloping landscapes. Because rain and other precipitation outpace evaporation in regions that have inadequate drainage, the water collects. Certain plant species, especially sphagnum moss, invade the area, and because of the high acid content, the dying vegetation does not decay. It accumulates over thousands of years so that eventually, as in the case of the black spruce-covered bogs of northern Minnesota, the peat can reach 25 feet in depth. It is saturated with water most of the year and is less capable of acting as a sponge than other wetlands.

Northern peat bogs are threatened primarily by efforts to harvest their peat for fuel and by encroaching agriculture. The fact that many of these bogs are located in remote wilderness areas protects them to some extent. However, scientists warn that remoteness won't protect peatlands from global warming, which would cause the peat to decay.

The effect of global warming on the northern peat bogs and other wetlands is important because of the role these ecosystems play in the great biogeochemical processes that cycle carbon and nitrogen from the atmosphere through plant and animal life and back again. Such wetlands represent some of the most productive land areas on Earth due to their capability of transforming solar energy into the chemical riches of living plants. Photosynthesis pulls carbon dioxide out of the air, where too much of it can overheat the planet, and replaces it with the oxygen that people and animals need to survive. If global warming decays these enormous pads of biomass, the many tons of carbon that they hold captive will be released back into the atmosphere, where a feedback effect could be set up, making things worse. (See GLOBAL WARMING, page 156.)

The contributions of wetlands to global ecology are as varied as the kinds of wetlands that exist. Marshes near metropolitan areas, such as those along the New Jersey Turnpike between Newark and New York, serve critical roles in removing pollutants from the air. Nitrogen is drawn from the air by microorganisms that convert it to ammonia and oxygen to nourish plants. Wetland vegetation is especially efficient at capturing excess nitrates and ammonia compounds produced by industry and agriculture, breaking some of them into harmless nitrogen gas which it returns to the atmosphere, and trapping still others in the mud and sediments underneath. Actually, some wetlands seem to be complex regulatory systems that store chemical nutrients as well as water, and then let them go when they are needed nearby. And unlike farmland, wetlands are powerful biotic engines that feed themselves on the recycled nutrients of decaying matter.

Swamps, marshes, river floodplains, and coastal wetlands are also like giant sponges, soaking up the excesses of floods and storing water as insurance against drought. They soften the blow of storms and high tides on coastal habitats, including human communities, and are natural filters that clean pollutants out of the water and recharge groundwaters.

Some marshes, like any of the prairie pothole marshes in Canada and the upper midwestern United States, are seasonal. They fill up with water from spring snow melt and rains and provide a summer home and breeding ground for about half of North America's ducks and other waterfowl. Marshes along the four major flyways provide rest stops for migrating birds, whose populations decline when droughts or intentional draining diminish these wetlands.

The important ecological role wetlands play wasn't understood until the 1960s. The practice of draining and filling America's wetlands began soon after Europeans arrived here. In 1764 George Washington helped to drain thousands of acres of the Great Dismal Swamp in Virginia for agricultural plans that failed. In the mid-1800s the federal government turned over 65 million acres to reclamation by the states. By means of these and many other reclamation projects, the 221 million acres of wetlands that existed when Europeans arrived have been reduced by more than half, and about 300,000 more acres are lost each year. Some states have lost more than 90 percent of their wetlands.

In the past few decades, the government has attempted to stop this trend. The Clean Water Act of 1972 assigned to the Army Corps of Engineers the task of issuing permits for filling wetlands; they can be overruled by the Environmental Protection Agency (EPA), the organization that is charged with providing guidelines for the protection of wetlands. In 1985, the U.S. Congress passed the Swampbuster bill, which withheld crop subsidies from farmers who drained wetlands. Enforcement of the Swampbuster regulations has been inconsistent. County agricultural agents and state agencies, as well as the U.S. Fish and Wildlife Service and the Soil Conservation Service, all have jurisdiction over different pieces of the wetlands puzzle.

In 1989, the "Federal Manual for Identifying and Delineating Jurisdictional Wetlands" was devised to coordinate all these efforts and set clear criteria for identifying wetlands, It quickly became controversial, however, and is symbolic of the conflicts that characterize wetland management. Environmental scientists find themselves at loggerheads with those interests—farming, building, and mining—that clearly could benefit from the freeing up of additional wetlands. Scientists point out that even under the current guidelines, the U.S. loses about 300,000 acres of wetlands per year. They warn that such losses will have long range environmental consequences and that any further deregulation of wetlands should be carefully avoided. Wetlands perform valuable buffering and filtering functions in the environment and are home to many species. Their loss, environmental scientists charge, will ultimately be ours, as well.

CONSTRUCTED WETLANDS

The capability of wetlands to remove wastes and toxins from water is serving as a model for human-made versions that use plants, microbes, and fish to convert waste into clean water.

Conventional methods, which treat millions of gallons of sewage each day in cities across the country, use concrete aeration tanks filled with microbes that consume organic substances; chemicals are used to purify the water and filters screen out viruses. This method produces tons of sludge, loaded with chemicals and heavy metals, that must be burned or buried in landfills.

Constructed wetlands, their proponents say, produce nothing more than clean water, plants, and fish. One such system employs greenhouses equipped with aquariums that are planted with water hyacinths and other water-loving plants. Algae and microorganisms living among the pebbles and roots at the bottom of the tank reduce the waste to nutrients that are used by the plants. Another tank containing an artificial marsh filters the water and converts nitrates into nitrogen gas. In yet another tank, fish gobble up the remaining microorganisms as the water is pumped through, and a second artificial marsh does the final filtering job in a system that mimics the efficiency of nature's own wetland filters.

In Europe and North America, reed beds or cattail marshes have been constructed to treat waste. The reeds soak up oxygen and transport it down their stems to the bacteria living in the water and mud. Meanwhile, the microbes, well supplied with the oxygen they need to stay alive, busy themselves breaking down the organic wastes in the water.

In the southern United States, there has been increasing interest in the benefits of constructed wetlands as a solution to household waste. Such systems make sense in states like New Mexico, where over 50 percent of the population lives in rural areas far from municipal sewerage systems. These bosques, as they are called, are maintenance free and succeed in areas where septic tanks and leach fields don't work because of clay soil and high water tables. A layer of odorless, composted leaves on the reed-bed surface does a good job of digesting the smelly gases that are associated with waste treatment, and so it isn't uncommon to see bosques in backyards. In an arid land like New Mexico, these islands of reeds, flowering plants, fish, and other wildlife are attractive solutions to the home wastewater problem.

Another alternative to traditional waste treatment is being implemented on a small scale in Ireland and Minnesota. There, researchers have constructed wastewater treatment systems from peat, a natural filter made of plant material that has accumulated in bogs for thousands of years. Both Ireland and northern Minnesota have some of the largest peat deposits in the world. Because of peat's ability to absorb bacteria, oils, heavy metals, and other industrial wastes, big chemical companies have shown interest in such experiments and funded research. Peat filters are capable of pulling mercury, copper, zinc, and other metals from water so that they can be recycled. In Minnesota, some individual homes and resorts are using peat filters to treat their wastes.

While the number of constructed wetlands that are treating wastewater on a small scale continues to grow, some engineers and biologists are skeptical that these systems could handle the massive amounts of sewage produced by a city each day. Supporters of what may ultimately be a healthier alternative to the current system suggest that the key lies in having smaller systems closer to home, rather than the giant centralized waste-treatment facilities now in use. After all, who would object to having small cattail marshes full of song birds, or greenhouses full of sweet-smelling flowers in the neighborhood?

BIODEGRADABILITY

Biodegradability is just a polite word for a harsh reality—every living thing is destined to decay. The agents of this destruction are Earth's all-important clean-up crew—the decomposers. These lowly members of the food chain, mainly bacteria and fungi, break down more complex organisms into the simple molecules that are the chemical ingredients for new life.

In this scheme, plants are the producers—transforming solar energy into food for the rest of the world. Primary consumers or herbivores eat some of these plants, but not nearly enough to keep Earth from overflowing with vegetable matter. Some of the herbivores are eaten by secondary consumers—the carnivores, scavengers and parasites. All the rest, including most of the plants, eventually die. If the molecules of water, carbon, nitrogen, phosphorus and other materials tied up in all these plants and animals were not recycled by the decomposers, life on Earth wouldn't last very long.

Given a chance, all materials derived from living things will biodegrade. But humans interfere with this process in a number of important ways. We bury most of our garbage in tightly packed, earth-covered landfills that block out the oxygen that decomposer bacteria need to do their work. Of the nearly one-half million tons of garbage that Americans throw out everyday, the biggest portion is paper—an easily biodegradable substance that too can last virtually forever in a landfill. Newspapers dug out of 1950s landfills are still readable!

We also turn organic and otherwise biodegradable compounds into plastic—a material made from long, complex molecules called synthetic polymers. The decomposers don't know what to do with these substances, and as a result, they too can last virtually forever. By adding biodegradable substances like cornstarch to their trash bag recipes, manufacturers have attempted to make bags that *will* decompose, at least partially, under the right conditions. Unfortunately, most of them end up with the newspapers in the air-tight landfill.

Taking a cue from nature, people are learning how to recycle things that would otherwise be buried. Glass, aluminum cans, newspapers, and some types of plastic are now routinely recycled in many parts of the United States. But most of these materials make their way into landfills and the environment.

GAIA HYPOTHESIS

Earth, the beautiful, rose up,
Broad-bosomed, she that is the steadfast base
Of all things. . . .

According to classicist Edith Hamilton, that is how the Greek poet Hesiod described the birth of our planet, which his civilization called Gaia, the mother-of-us-all. The Greeks assigned a kind of detached personality to her, a being who was hard as rock and yet dynamic and alive.

Throughout history, most scientists have taken a more prosaic view. But over the years a few have proposed that Earth is, in fact, a dynamic organism, not just a rock table on which to support the real activities of life. The most recent proponent of this idea is British scientist James Lovelock who, while working on a NASA project assessing the feasibility of life on other planets, was forced to think about the differences between Earth and her dead neighbor Mars.

The problems involved in seeding life on Mars inspired Lovelock to examine the delicate shell of life that embraces Earth. The challenge, he found, is not merely getting earthlike life to grow on Mars, which is a formidable problem by itself; it involves duplicating the self-regulating mechanism that allows the planet to roll with the environmental punches of geological disasters and collisions with meteors, comets, and other space debris, as well as the insults of human-made pollution and destruction.

Lovelock and other proponents of the Gaia hypothesis suggest that Earth, in cooperation with the life it shelters, is even able to repair itself when bad things happen. Since life first emerged some 3.5 billion years ago, asteroids, comets, and other chunks of rock have struck Earth, wiping out many, but not all, living things.

For example, scientists think some such collision wiped out the dinosaurs 66 million years ago. The impact threw tons of dust and ash into the air, blocking off the sun and chilling the planet. Many species of plants and animals died—but most kinds of life on Earth survived. In fact, scientists think the exinction of the big dinosaurs made room for mammals, including humans, to evolve and thrive.

Earth's apparent ability to regulate its atmosphere, some scientists say, is another reason to believe in Gaia. During its infancy, the planet was probably blanketed by a thick coat of atmospheric CO_2, creating a greenhouse effect that compensated for the weaker heating capacity of the early sun. The blanket of CO_2 trapped the sun's warmth inside the atmosphere, warming the planet. As the sun's power grew, the greenhouse effect was gradually reduced by living things, which consumed excess CO_2 and thus maintained a habitable environment in which to flourish.

Lovelock has written that Earth truly is Gaia, a self-regulating, self-repairing organism, not just a template for the creation and continuation of life. All the living things on the planet regulate Gaia's evolution by their growth and interaction and, along with the oceans and the atmosphere and the great geochemical cycles that serve as Gaia's circulatory system, help her to respond to challenges, in the same way that any living thing responds.

While Gaia seems equipped to survive tremendous challenges, Lovelock notes, she may be severely stressed by the changes that humans have brought about in the giant feedback system that she uses to remain stable. Even so, the worst challenges, the scientist reminds us, have not destroyed life on Earth as it must have been destroyed or precluded on other planets in our solar system. Gaia may survive the worst that humans can do to her, Lovelock says. It's just that humans and many other species may not.

IV

TECHNOLOGY AND THE INTELLIGENT MACHINE

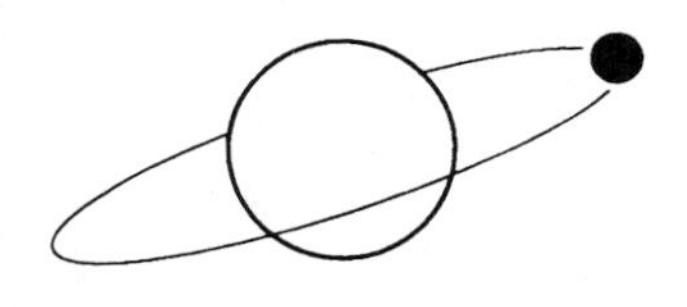

DURING THE TWENTIETH CENTURY—especially since the 1940s—technology has raced forward at such a terrific pace that most people can't keep track of it. Scientists' wildest fantasies have come true. And yet, when researchers explore what seems to be the distant frontiers of knowledge, people ask, "What good is this? How will this ever help me?"

For example, fifty years ago, physicists studying the physical properties of solid materials must have seemed like incredible eggheads working on the most esoteric stuff. But the products of that research are everywhere in our daily lives. Solid-state physics gave us transistors, semiconductors, and integrated circuits, all components of the microelectronics revolution. Without them we would not have many of the things that we use everyday, including transistor radios, pocket calculators, and computers. In fact, nearly everything we use has some connection with microelectronics. They are in our cars, cash registers, pacemakers, toys, washing machines, televisions, refrigerators. The list goes on and on.

Such things as levitating trains, lasers, moving pictures of babies in the womb, robots, and intelligent machines were the stuff of science fiction not so long ago. Who would have thought that we'd use lasers in our stereos, stores, and hospitals? Or that computers would be diagnosing illnesses, approving loans, or coordinating the movement of hundreds of battleships and airplanes? Or that trains would be flying along on pads of magnetic resistance?

Research doesn't always produce practical tools and entertainments, though. It's important for its own sake, and it helps us better to understand ourselves, our world, and our universe, as well as sometimes giving us technologies that demand decisions be made. How much are we willing to spend for high-tech transportation or medicine? How much of our lives should be managed by computers, and what happens to our privacy when personal information is only a keystroke away?

Because of complicated jargon and the pace of change, keeping up with technology is more difficult, but it's in our best interest to do so. We need to use our imaginations, ask questions, and pay attention, because technology has the power to change our lives forever.

BUCKYBALLS

Researchers all over the world are playing with Buckyballs, a new form of pure carbon that was discovered by chemists in 1985. Microscopic soccer ball–shaped crystals of Buckminsterfullerene just over a nanometer—or billionth of a meter—in diameter, Buckyballs are members of a newly discovered class of molecules called fullerenes. They are named after R. Buckminster Fuller (1895–1983), the architect who designed the soccer ball–like geodesic dome.

Pure carbon molecules like Buckyballs are a component of the dust that drifts between the stars. Scientists think that carbon-filled stars produce the tiny molecules and blow them into surrounding space. Buckyballs, which often contain 60 atoms of carbon, have some strange qualities. Under extreme pressure they can form a substance that is harder than diamonds, but when the pressure is released, they pop back to perfectly round, slippery spheres that can be lined up evenly on surfaces like little beads. Diamonds, the

Buckyballs

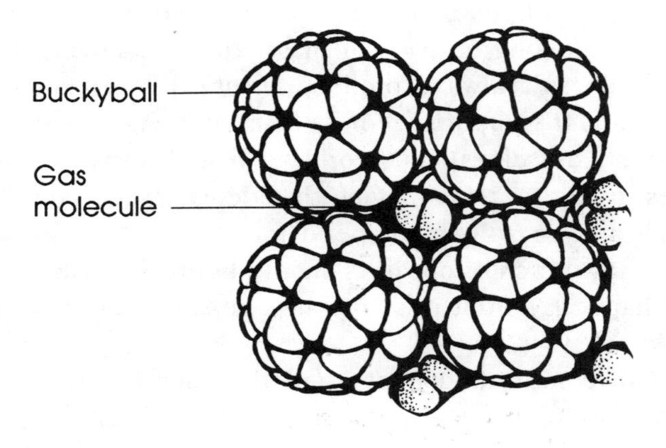

hardest known substance, and graphite, the soft material in lead pencils, are the only other known forms of pure carbon.

Physicists learned to produce Buckyballs in quantity in 1990. Now, scientists are excitedly looking for down-to-earth uses for C-60, which is produced by vaporizing graphite with laser blasts. When the vaporized carbon cools, it forms into spherical carbon molecules, some of which are Buckyballs. Apparently the graphite, which exists in layers of hexagonal carbon molecules, is broken apart into fragments that look like scraps of chicken wire. Dangling at their edges are carbons bonds looking for something to grab onto. The carbon atoms shuffle around to make pentagons, forming new bonds that pull it into the familiar shape of the soccer ball.

It's been said that C-60's unusual shape defied discovery until one night a frustrated researcher went home and, after a few beers, worked it out on the kitchen floor, using scraps of paper in the shapes of pentagons and hexagons. The structure distributes stress evenly, making Buckyballs very stable and strong.

Other fullerenes are probably present in any sooty smoke, such as a candle flame. But researchers think that Buckyballs are less likely to be there because the hydrogen in air jumps and grabs the hands of those dangling bonds. The curling effect that creates the soccer ball–shaped molecule probably curls the carbon atoms into long spirals.

Those dangling bonds have proven useful to physicists who are trying to "grow" thin diamond films on silicon surfaces. They have found a way to prepare a silicon surface with shreds of Buckyballs that grab onto other carbon atoms and begin the process of forming diamond films. Besides being one of the hardest known substances, diamond is a tough, slippery conductor of heat and current. It would be very useful in electronics, where it could shed heat and withstand temperatures that other materials used in semiconductors could not. As a coating for mechanical parts, it could withstand greater extremes of heat and friction.

The cagelike structure of fullerenes also suggested to researchers that it might be possible to insert another kind of atom inside the cage before the dangling carbon bonds latch together. In late 1991 they succeeded in doing that very thing: By blasting other substances with lasers in the presence of graphite and fullerenes, they were able to capture metal atoms inside the fullerene cages. In combining the properties of fullerenes with those of other substances, scientists may be able to make materials with amazing capabilities of lubricating or conducting electricity.

The most promising use so far for the Buckyball may be its ability to turn other materials into superconductors. A superconductor is a substance that has no resistance to an electric current. If you were to feed a superconducting circuit just a bit of electricity and then remove the source of the current, the circuit would pass the "juice" around and around indefinitely. A lot of electricity is usually wasted as it passes through conventional wires and circuits because the materials resist the flow of current, creating heat. Superconductors might save billions of energy dollars every year. (See SUPERCONDUCTIVITY, page 191.)

For the time being, superconductors are impractical for widespread use because they work only when chilled down to temperatures near absolute zero, or −459°F. But by making a thin film combining Buckyballs with a metal like rubidium and with other substances such as cesium and potassium, scientists are able to make a superconductor that functions at higher temperatures, something certain ceramic materials can already do. Apparently, the Buckyballs share electrons with the metal atoms, forming a kind of bucket brigade that passes electricity along paths that would not otherwise exist.

Experiments have also shown that Buckyballs can act as gates, shutting off the flow of current under certain conditions. Such substances, therefore, would be useful as sensors that would turn current on or off under specific conditions. The usages for Buckyballs constitute a research area that is ripe with possibilities.

ELECTROMAGNETIC FIELDS

Magnetism and electricity are two aspects of the same force—electromagnetism—that keep electrons spinning around the nuclei of atoms and causes atoms to cling together in molecules. Electromagnetism is also the power that propagates light waves and all the other components of the electromagnetic spectrum. A changing magnetic field induces an electric field and a changing electrical field induces a magnetic field. In this way the two travel along locked together but dancing perpendicular to each other, and to the direction of the motion of the wave.

The magnetic fields around a magnet or moving electrical charge are lines of force that connect its north and south poles. If you scatter some iron filings on a piece of paper and hold a magnet underneath it, you can see that the filings arrange themselves along these magnetic lines of force. In the same way, you can also observe the attraction between the north and south poles of two magnets and the repulsion between like poles.

What this little experiment does not reveal is the continuation of the magnetic lines inside the bar magnet where they form closed loops. The magnetic fields around and within iron magnets come from the electrons in its atoms, the source of the moving electrical charge.

The same lines form closed loops around a wire that is carrying a current. A wire wound into a coil and carrying a current will have lines of force coursing through the center of the coil and around the outside in closed loops. It exhibits the same kind of magnetic field as a bar magnet.

Every electric charge has an electric field that varies in intensity depending upon the strength of the charge. Lines of force run away from negative charges and towards positive charges, so that unlike charges are attracted to each other. And oppositely charged parallel plates (a uniform field) have parallel lines of force. The density of the lines indicate the strength of the field.

The connection between electricity and magnetism was first noticed in 1820 by Danish physicist Hans Christian Oersted (1777–1851). While teaching a class one day he flipped a switch that allowed current to pass through a wire, causing the needle in a nearby compass to jump. After studying the situation, Oersted realized that the electricity flowing through the wire had created a magnetic field that had affected the compass.

Stimulated by Oersted's research, French physicist André Marie Amperè (1775–1836) found that current flowing in the same direc-

Magnetic Fields

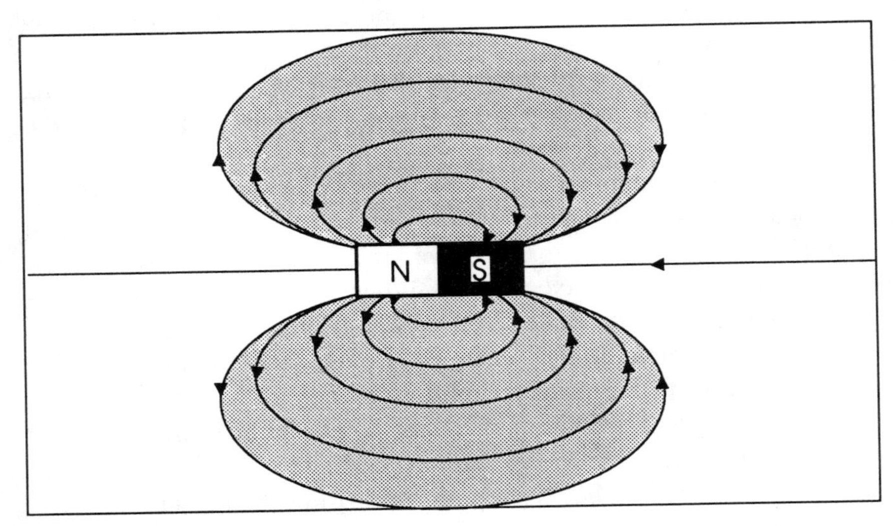

The three-dimensional magnetic field that surrounds a single bar magnet has lines of force coursing from the north to the south pole.

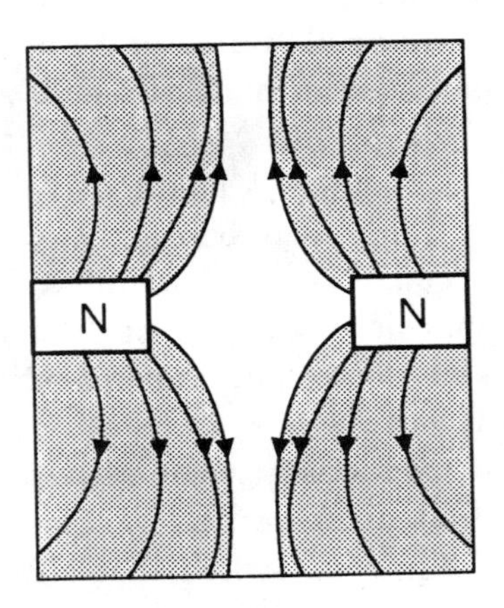

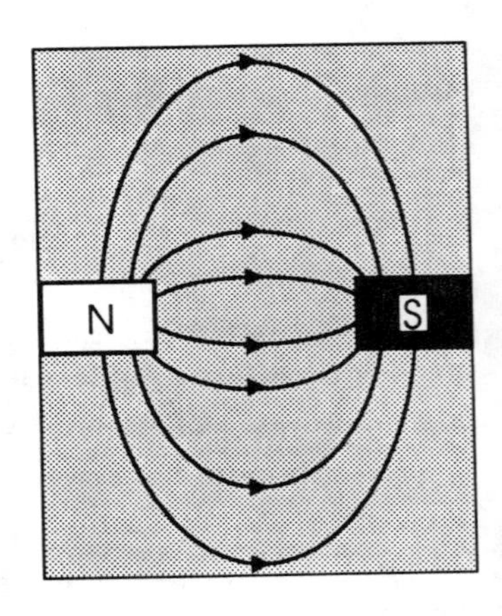

Like charges repel each other and opposite charges attract, as illustrated by the drawings above. In the first illustration (left), the like poles repel each other, creating a neutral pocket where there are no lines of force. However, the two opposite poles (right) are attracted to each other and are linked together by lines of force.

Electric Fields

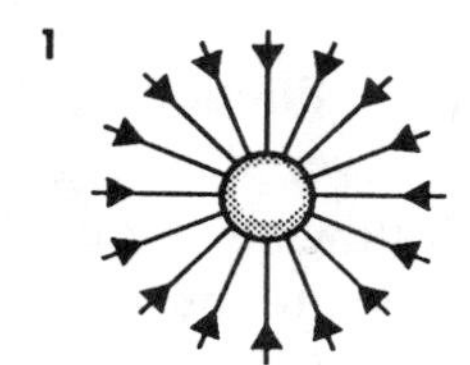

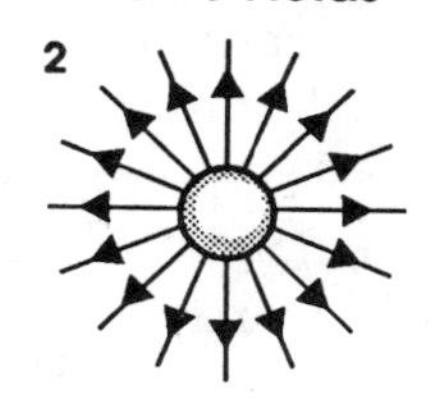

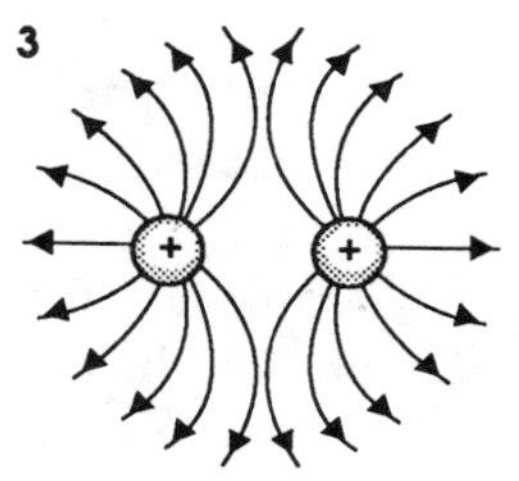

An electric field is represented by lines in space, and the stronger the field, the closer these lines are to each other. The arrows show the direction of the field, which flows toward a positive charge (1) and away from a negative charge (2). While they appear here in two dimensions, the lines of force actually form a three-dimensional pattern. Like charges repel (3) and unlike charges attract (4). Oppositely charged parallel plates form a uniform field in which the lines are parallel (5).

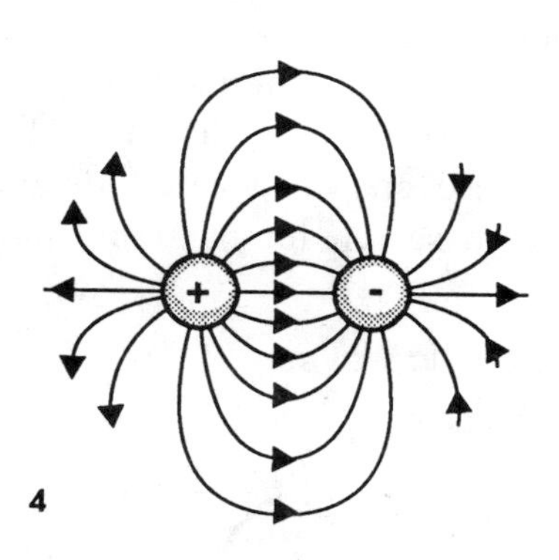

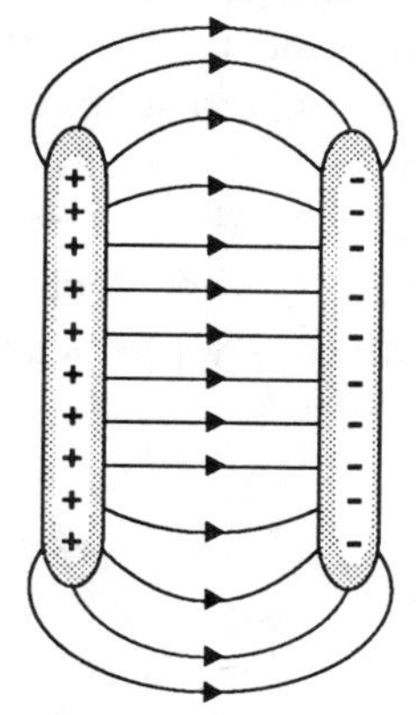

tion through two parallel wires caused an attraction between them, while current flowing in opposite directions caused repulsion. He proposed that electricity is actually the source of magnetism and showed that a coil of wire with current running through it behaved just like a bar magnet.

British physicist Michael Faraday (1791–1867) reasoned that if current flowing through a coiled wire could produce magnetic fields, why couldn't a magnetic field induce a current of electricity in a coil of wire? His experiments, in which a magnet was moved back and forth in or near a coil of wire, showed that a magnetic field could generate an electrical current without any other source of voltage feeding the wire. It is this electromagnetic induction that forms the basis for the electrical power that we use everyday.

By rotating a wire in a magnetic field, the current flowing through the wire passes in one direction for a time and then in the other direction for the same amount of time. This is called alternating current, and it's the principle behind the rotating generators that provide our electricity. A transformer uses the current passing through one wire loop to induce current flow in another loop. Transformers are used to increase or decrease voltage. The voltage that comes from the generator at the power plant is increased by a transformer to travel over power lines and decreased by another transformer before entering the house or building where it's used.

Scottish physicist James Clerk Maxwell (1831–79) devised four equations to describe what Faraday had figured out about electromagnetism and together they encompass every aspect of the inseparable relationship between electricity and magnetism. Maxwell's equations state that like charges repel each other and opposites attract, electrical current produces magnetic fields, magnetic fields can induce electrical current, and there are no isolated magnetic poles. (Although physicists continue to search for magnetic mono-

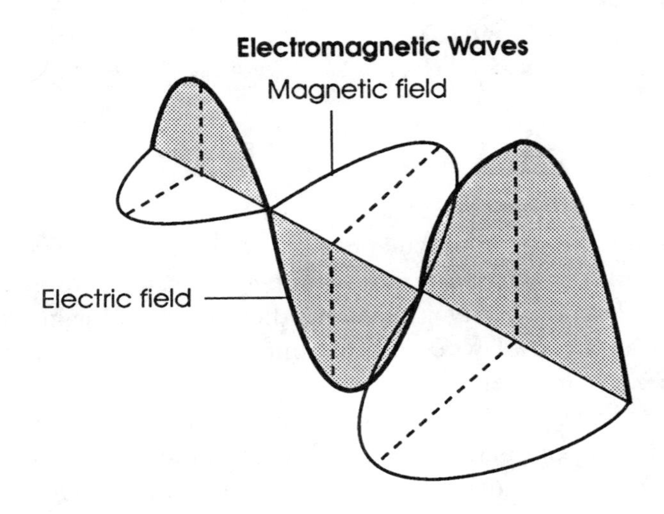

poles, massive particles left over from the Big Bang that a have single pole of magnetism, their existence has yet to be proven.)

SUPERCONDUCTIVITY

A superconductor is a substance that has no resistance to an electric current. If you were to feed a superconducting circuit just a bit of electricity and then remove the source of the current, the circuit would pass the "juice" around and around indefinitely.

A lot of electrical energy is wasted as it passes through conventional wires and circuits because the materials resist the flow of current. The thermal vibrations of atoms get in the way of electrons trying to flow through the conductor. As a result, the electrons bash into the atoms and transfer their energy as heat. But when some conductors are chilled below a certain temperature, the thermal vibrations of atoms are stilled and electrons can flow through the material unimpeded.

Scientists have known about this phenomenon since 1911 when Dutch physicist Heike Kamerlingh Onnes (1853–1926) observed it in chilled mercury. Experiments with various materials continued, but the problem was always the same. Materials could not act as superconductors unless they were chilled to near 0 K (Kelvin is a temperature scale that measures the degrees above absolute zero, the temperature at which all molecular activity ceases. It is equivalent to −459 F or −273.15° C.)

Even so, researchers succeeded in developing superconductors that could carry electrical currents far greater than a regular metal conductor would tolerate. Powerful electromagnets were possible, but they had to be cooled with liquid helium, an expensive and fussy substance. (Powerful, water-cooled electromagnets already existed before the advent of superconductors and are still in use, but tremendous amounts of energy and water are required to operate them.) Then in 1986 when K. Alex Müller and J. Georg Bednorz at the IBM Zurich Research Laboratory discovered a ceramic that was superconducting at 30K, superconductivity research really heated up. Experiments quickly led to the development of ceramic materials that were superconducting at temperatures as high as 120K so that liquid nitrogen—easier to manage and much cheaper—would do just fine.

Although these superconductors function at what we would still consider very low temperatures, the technology is referred to as high-temperature superconductivity. The ultimate goal, however, is the development of room-temperature superconductors.

Despite advances in electronics that have made components smaller and smaller, the heat produced by resistance limits how closely they can be packed together. Connecting leads made of high-temperature superconductors could solve that problem.

Researchers also have used superconducting films, with an insulating layer sandwiched between them, to create tiny ultrafast, supersensitive switches called Josephson junctions.

In the meantime, superconductors, although expensive, are at work in hospitals and laboratories in powerful imaging machines that use superconducting magnets. The magnets in these magnetic resonance imaging (MRI) machines for the hydrogen atoms in the body to snap to attention along magnetic lines. Then the atoms are bathed with radio waves that cause the atoms to change position. When those radio waves are turned off, the hydrogen atoms blink out a chorus of tiny signals that are collected by a computer and converted into an image. MRI is also used in industry to detect otherwise undetectable flaws in metal. (See MAGNETIC RESONANCE IMAGING, page 223.)

Superconductive magnets are also used in experimental fusion reactors to create an invisible magnetic bottle that isolates the heat and energy of nuclear reactions. Fusion reactions take place at temperatures that would vaporize any known material.

Superconductivity may some day be used to create immense power systems that are able to deliver energy from remote sources to cities thousands of miles away. Right now, because of the resistance of the conductors used in power lines, millions of energy dollars are simply wasted. Power companies may also be able to store extra energy by circulating it in superconducting loops until the peak hours when it is needed.

Finally, superconductors may one day help us get from here to there. Levitating trains that are pulled along at high speeds by the attraction of superconducting magnets could be the answer to many modern transportation woes.

MAGLEV (MAGNETIC LEVITATION)

Most school children have experimented with magnets and observed the attraction between opposite poles and the invisible pad of repulsion between like poles. Scientists are using the same principles to develop a new kind of transportation: They employ the push/pull properties of magnets to make a train that flies at high speeds on a pad of air above a rail or trough. The technique is called maglev, short for magnetic levitation.

There are two basic types of maglev, electromagnetic suspension and electrodynamic suspension. In the electromagnetic suspension system developed in Germany, electromagnets are embedded in the section of a streamlined train that wraps under the rails. The magnets are pulled up toward the rails, but their current is adjusted to maintain a space so that they don't touch. This kind of maglev is propelled by attractive force between the steel rails and the magnets on the train.

Electrodynamic suspension, tested in both Japan and the United States, uses repulsive force to move the train. The vehicle rides in a U-shaped guideway that is embedded with copper coils alongside and below the train. The train itself is fitted with superconducting magnets that are cooled by liquid helium. The polarity of the coils in the walls is alternated so that the train is pulled from in front and pushed from behind by a traveling magnetic field, all the while being suspended in air two to six inches away from the surface of the roadway. Because the maglev doesn't touch the track, no friction with the road slows down the train. However, there is some air friction, especially at high speeds. One such device being tested in Japan can go more than 300 miles per hour. Such trains are also fitted with regular wheels and tires, which can be used for rolling into stations or dropping onto should the superconducting magnets fail.

Since maglevs don't use fossil fuels, governments are interested in these high-speed trains as solutions to pollution and energy problems, as well as airport and highway congestion. Maglev's proponents say electric trains will help wean transportation away from fossil fuels, and they are ideally suited for short connections between cities like Washington, D.C., and Boston, Pittsburgh, and Philadelphia or between Los Angeles and Las Vegas. Such short hops would be faster on a maglev, and the elimination of commuter flights would lighten the load at airports. The first commercial maglev is planned to go into operation in 1995 in Orlando, Florida, connecting the airport with resorts in the area. Maglev guideway might be built on the median strips of interstate highways, and then rise above traffic, parking, and walkways in busy cities.

One of the chief drawbacks to maglev is the difficulty in shielding passengers from the strong magnetic fields that are emitted by the superconducting magnets. Although no studies have offered conclusive proof, exposure to magnetic fields has been associated with cancer and other medical problems. Strong magnetic fields may also interfere with pacemakers, prosthetic devices, and the electronics in cars and other machines.

Nevertheless, these speedy land vehicles have been considered much safer than air or automobile travel, based on the performance of high-speed trains in Europe and Japan. However, Japan suffered a disappointing setback when its experimental maglev caught fire after a tire had blown out while the train was operating at a low speed. The U-shaped track confined the flames and completely gutted the vehicle.

Cost is also an important consideration. Guideway construction could cost from $8 million to $60 million per mile, according to some studies. U.S. researchers began working on maglev during the 1970s, but government showed little interest in developing high-speed train travel until as recently as 1991.

ELF MAGNETIC FIELDS

In this electronic age we are surrounded by seemingly innocent low-frequency electrical fields from small kitchen appliances, electric blankets, computer terminals, and everything else that is powered by electric current moving through wire. Since the late 1970s, however, studies have shown a link between the electromagnetic fields generated by power lines and the incidence of cancer, particularly childhood leukemia.

Two Denver studies examined the relationship between childhood leukemia deaths and the location of electrical distribution lines in neighborhoods. Those children who lived near the lines were twice as likely to develop leukemia as the control group of children who did not.

More recently, researchers have examined the effect of extremely low frequency (ELF) emissions—below 300 hertz—the kind with which we surround ourselves in our electronic homes and offices. The 60 hertz, or cycle per second, electrical current that powers most American homes has long been thought harmless. Now some scientists have raised doubts.

Laboratory experiments on living cells and animals have suggested a relationship between magnetic fields and the activity of DNA and RNA. Cells exposed to electromagnetic fields increased by up to 400 percent the rate at which RNA transcribed, or copied, DNA, the genetic blueprints for the cell and the larger organism. The research concentrated on genes that, when turned on by some stimulus, can trigger the wild cell division that happens in cancer, in this case leukemia. Other studies found that exposure to electromagnetic fields altered the activity of chemicals that are associated with the development of cancer.

ELF exposure has also been implicated in birth defects in animals and in the occurrence of miscarriage among women exposed to emissions from video display terminals (VDT) during the first trimester of pregnancy. Those who used VDTs for more than twenty hours per week were twice as likely to suffer miscarriages, studies found.

The concern about emissions from VDTs has become a work-safety issue. Strict standards controlling such exposure are being adopted by several European countries. In the United States, people have sued their employers over damages incurred from the use of VDTs and other components of computer systems, and the first legislation aimed at controlling VDT exposure was passed in San Francisco in 1990.

While most experiments have shown a statistical association between ELF emissions and cancer, rather than a cause-and-effect relationship, the evidence has been strong enough for the Environmental Protection Agency to issue a 1990 report stating that it is "biologically plausible" that electromagnetic fields could cause cancer and that the available data suggests a link but has not yet proven it. A paper issued by the Congressional Office of Technology Assessment concluded that ELF emissions are a source of concern and can affect the way cells function but may not present a significant risk to public health.

Nevertheless, many scientists consider the evidence associating magnetic fields with disease to be inconclusive.

MICROWAVES

Our most familiar association with microwaves is in the kitchen. Practically everyone has seen or used a microwave oven at some point, even if they don't have one in their home. But before microwaves were used to cook food, they were being used to detect aircraft, carry telegraph messages, and explore the surfaces of distant planets.

Actually, microwaves are emitted at some level by most objects—unless they are so cold that all molecular activity has ceased (and nothing in nature is that cold). Furthermore, our universe is bathed with a uniform wash of microwave radiation that scientists believe is leftover from the Big Bang.

Microwaves are high frequency waves that occupy a spot on the electromagnetic spectrum about midway between radio waves and light waves. (See ELECTROMAGNETIC SPECTRUM in Part I, page 10.) They have some of the qualities of both and are also used to carry communications and to help us see things. They travel at 186,000 miles per second—the speed of light as well as all other kinds of electromagnetic radiation. And like light beams, microwave beams travel in a straight line and are reflected by solid objects.

Unlike radio waves, microwaves do not bounce off the ionosphere, the layer of the atmosphere that is filled with charged particles. As a result, communications systems using microwaves depend upon line of sight transmission. As early as 1945 they were employed to carry telegraph messages, vastly improving the system in use at the time because by using microwaves, the same frequency can be used to carry thousands of messages on narrowly focused beams. However, because of the line of sight problem, they require relay towers about every 30 miles to reflect and boost the microwave signal. On the plus side, microwaves are not affected by atmospheric conditions, like lightning, that interfere with other radio signals.

Today, satellites serve as relay towers in space, bouncing signals from one part of the world down to other points on the globe. All sorts of communications, including telephone calls, television, and radio are carried on microwaves and relayed by satellites.

Radar, developed during World War II, uses microwaves to track airplanes. During the 1940s a Hungarian scientist, Zoltan Lajos Bay, was able to bounce microwaves off the moon and detect the reflection of those waves. Then in the 1960s astronomers were able to bounce microwaves off of Venus and Mercury. Since they knew how fast the microwaves traveled (186,000 miles per second), and how long it took them to get there, they were able to more precisely calculate the distances to those planets. Microwaves were also used to penetrate the thick clouds of Venus and determine the speed and direction of its spin, which is puzzlingly counterclockwise.

Down on Earth, one of the most useful task for microwaves is cooking; they excite the food's loosely bound atoms and molecules, causing them to spin faster or in different directions. All this movement makes the food hot. The more tightly bound atoms and molecules in glass, ceramics and some plastic dishes are not affected by microwaves, so they stay cool. Microwaves will not pass through metal, however, and containers made from it are useless in microwave ovens.

In a microwave oven, the microwaves are generated by a magnetron, a device similar to the picture tube in a television. It shoots a beam of electrons that oscillates billions of times per second, making microwaves. The beam strikes the spinning blades of the fan, which scatters the microwaves all over the inside of the oven, penetrating the dish and the food.

DOLBY

Early cassette tapes were hissy, noisy things. The subtleties of music were lost to background noise, especially during passages that had soft, high-frequency sounds. (See ELECTROMAGNETIC FIELDS, page 186.) An American engineer, Ray M. Dolby (1933–) invented a way to eliminate that noise.

Recording tape is a long plastic strip that is coated on one side with magnetic particles. Tape recorders convert electrical signals into magnetic information on the tape. The soundwaves from instruments and voices are converted by microphones into electric signals. They, in turn, are converted into magnetic fields that change the pattern of the particles on the tape. When it's played back, the magnetic message on the tape is converted back into an electrical signal that is amplified by the machine and changed back into sound waves by speakers or earphones.

Noise caused by the equipment or the magnetic particles, is picked up by the recording along with the desirable sounds. Even the best magnetic tape and machines introduce some hiss into the recorded sound, especially at slow speeds. Cassettes contain especially narrow tape, thus limiting the area available for the magnetic message, and they play at a relatively slow speed. Both contribute to tape hiss.

Ray Dolby devised a way to boost the sound frequencies that we want to hear, while reducing the hiss. The Dolby mechanism includes a circuit that encodes or amplifies the frequencies of sound that are especially quiet before they are recorded onto the tape. The desirable signals are now much stronger than the hiss or background noise. When the tape is played back, the Dolby circuitry decodes the signal, bringing the amplified quieter passages back to their natural level in relationship to the rest of the desirable sound. But the background hiss, that wasn't amplified along with the quiet passages, is significantly reduced in the decoding process.

Early Dolby systems were able to reduce noise by 10 decibles. Decibles (dB) are units for measuring the intensity of sound on a scale that increases geometrically rather than arithmetically. Twenty dB are ten times greater than ten dB. Newer Dolby systems claim to reduce noise by over twenty dB, making hiss virtually inaudible and competing with the clean sound of digital recordings. (See DIGITAL/ANALOG below.)

DIGITAL/ANALOG

A child learning to count uses fingers to add things up. Digit, as in digital recording or clock, gets its name from *digitus*, the Latin word for finger. All the uses of digital technology are just an extension of that simple idea. Digital technology is used to measure, in discrete parcels or numbers, such disparate things as time, sound, and information.

We're used to hearing about digital watches, computers, and recordings, but we seldom encounter references to analog devices even though we use them everyday. The clock on which we all learned to tell time—the "old-fashioned" kind with a face and hands—is an analog device. It displays time as a continuous series of angles and relationships. Conventional television and radio are analog messages carried by continuous wave forms. Audio tape recorders transfer the electrical signals of wave forms onto tape by magnetizing particles on its moving surface.

Digital recordings, on the other hand, sample, or take bites of, the amplitude of the wave form as many as 44,000 times per second. The intensity of sound is converted into binary digits—series of ones and zeros—that can be stored on magnetic tape and computer discs as spots of high and low magnetism and on compact discs as tiny pits that are read by lasers.

It's as though the waves were sliced into tiny pieces, each of which carries a speck of sound that can be stored separately. When all these specks are put back together, the message is remarkably faithful to the original sound—close to the limits of human hearing.

At the same time, digital recordings eliminate the noise or hiss that can only be diminished on analog recordings with noise reduction systems. (See DOLBY, page 198.)

COMPUTERS

The microelectronic device that has had the most profound effect on our lives and our society is the computer. Its historic origins are the computational devices that people invented to make problem solving easier. In the broadest sense, the computers that keep track of nearly every detail in modern life had their roots in the abacus, a device that uses strings of beads to solve arithmetic problems. It was invented thousands of years ago and is still in use in some parts of the world.

There are more recent precursors of the computer as well. French mathematician Blaise Pascal (1623–62) is credited with making the first mechanical calculator, and German philosopher and mathematician Gottfried Wilhelm von Leibniz (1646–1716) also invented a calculating machine and worked on the notion of breaking mathematical problems into a series of easier, smaller steps. In addition, Leibniz explored the usefulness of binary systems. Binary switching—alternations between yes and no, on and off—as we'll see later in this section, is the foundation of modern electronic computers.

Other contraptions that led ultimately to today's supercomputers were the inventions of a nineteenth-century eccentric, mathematician and inventor Charles Babbage (1792–1871). He envisioned two mechanical computing machines, elaborate steam-driven metallic devices that used interlocking geared cylinders and levers to solve problems. With the help of the British government he built a model of his Difference Engine, which was designed to calculate math tables, but before completing the full-sized version he turned his attention to another dream—a more complex device that he called an Analytical Engine. Designed to calculate and apply logic to problems, it used programs in the form of punched cards, similar to those that controlled the weaving patterns in fabric looms. It also had a memory and used the binary system.

But, alas, poor Babbage couldn't get funding for the construction of his calculating machine and spent the rest of his life trying to complete a computer that was hopelessly ahead of its time. A simpler version of the Analytical Engine was built in Sweden and won a gold medal at the Great Exhibition in Paris in 1855.

An American engineer, Herman Hollerith (1860–1929), designed a tabulating machine that was used in the 1890 U.S. census. It was the first calculating machine powered by electricity and the first to use Babbage's punch-card idea to input data. It proved to be ten times faster than counting census cards by hand and so captured the world's interest. Hollerith decided to start a company to manufacture his tabulating machines, and in 1924 it became the International Business Machines Corporation, or IBM.

These early inventions represent the beginning of a revolutionary idea—machines doing intellectual activity—that did not come to fruition until many years later when the first electronic computers were invented during the 1930s. Babbage would have loved them.

The first, big, working electronic computers were built in time to play a role in World War II. The Mark 1 was used to construct tables that would tell how to aim guns in order to hit targets accurately. It took into consideration such factors as distance, weather, and types of shells fired. It used 500 miles of wire and 3,300 electromechanical switches and was programmed with punched tape. The electric motor and metal shaft that drove the Mark 1 sounded "like a roomful of ladies knitting," one scientist said. It was quickly outpaced by ENIAC, the Electronic Numerical Integrator And Calculator built in 1946 to design the hydrogen bomb. ENIAC replaced switches with vacuum tubes.

ENIAC and its descendant UNIVAC represent the first of four generations in computer technology. In 1954 UNIVAC accurately predicted that Dwight D. Eisenhower would defeat Adlai Stevenson in the presidential election. The business world was slow to see the usefulness of computers, but eventually they turned over some of their bookkeeping tasks to "electronic brains," as they were called. Requiring elaborate cooling systems because of the heat-producing vacuum tubes, these giant computers ran hot, took days to program, and broke down frequently.

With the invention of transistors in 1947, computer technology took a giant step forward. Transistors are two hundred times smaller than vacuum tubes and capable of performing more funtions faster, more reliably, and without as much heat. They are switches made from layers of semiconductors, materials that transfer less electric current than a conductor like copper, silver, or aluminum, and more than an insulator like wood, glass, or plastic.

Silicon, the most commonly used semiconductor, is a brittle, crystalline element that, combined with other elements like oxygen and aluminum, is as common as clay and as plentiful as sand on the beach. In fact, next to oxygen, it's the most common element in nature. It's useful in electronics because it has a few electrons that are less tightly bound to their atoms and can be easily shaken loose to carry a limited amount of current. By "doping," or adding impurities, to silicon or some other semiconductor, one can alter its conductivity and endow it with negative (N) or positive (P) charges.

By sandwiching together semiconductors with different charges, manufacturers can build tiny electronic components that function as switches, amplifiers that boost weak electric signals into stronger ones, and rectifiers that convert alternating current (AC) from the outlet into the direct current (DC) that is needed by the circuits inside electronic devices.

When N and P semiconductors are sandwiched together, they set up walls of opposing charges at their boundaries. An electron, which has a negative charge, is attracted by the P, or positive layer, and repelled by the N, or negative layer. In this way, an electron is rushed across the wall and given a metaphoric kick in the pants for good measure, which contributes to its energy level.

When N and P are layered in such a way as to resemble a bologna

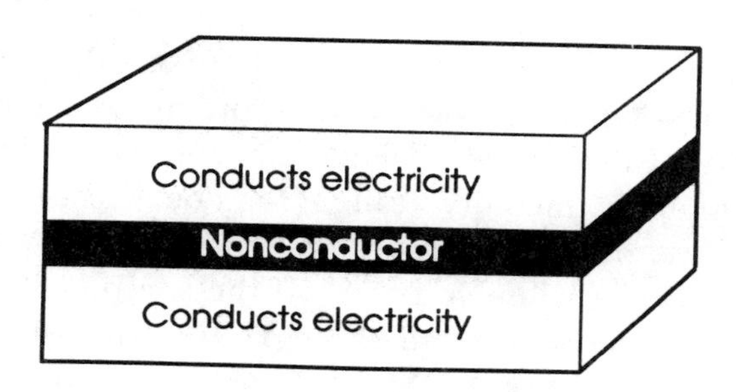

sandwich, the semiconductors become a switch. The middle layer, or bologna in the sandwich, is connected to a source of current. When it's turned on, the electron is ushered right through the sandwich. When the current is turned off, the electron can't pass. Changes in the voltage applied to the middle layer can also amplify the current passing through. Thousands of these tiny sandwiches can be built in fine layers on slices of silicon that are about the size of your thumbprint. Just one of these microchips does the work inside your pocket calculator.

The transistors and other devices inside computers control the flow of electricity by switching on and off millions of times per second. Word processing, math problems, and the complex streams of information that computers manage are reduced to on's and off's. One such switch to on or off is called a bit, which is short for binary digit. Eight bits are equal to a byte.

The binary number system, the language computers use to do their work, relies on combinations of 1 and 0 (or on and off) to represent numbers, letters, and symbols. The letter A, for example, is 01000001. Computers are able to manage thousands of bits in a fraction of a second. And the more transistors and other components on a chip, the faster the computer's switches can flip on and off, routing, storing, and retrieving information. A computer's ability to store information is measured in bytes and is usually stated in kilobytes (K), or thousands of bytes. For example, a machine with 128K has storage room for 128,000 bytes. Supercomputers can store millions of bytes.

All computers, from the simplest home model to the huge mainframe, have some variation on these basic components:

An input device, such as a keyboard, mouse, optical character reader, or speech recognition device; A central processing unit (CPU) that manipulates information according to a program or set of instructions. Programs are called software; the machine itself is called A temporary memory (called Random Access Memory, or RAM) and a permanent memory (called Read Only Memory, or ROM). RAM is lost when you turn off the computer, or if a power outage does it for you, but all those 1's and 0's can be saved on magnetic disks or tapes. ROM is special information that has to do with the operation of the computer and can't be erased;An output device, such as a monitor, printer, speech synthesizer, or modem (a device that sends data over the phone). All these extra pieces of hardware attached by cables to the computer are called peripherals.

Technological advances continue to make computers smaller, cheaper, faster and more powerful. While we increasingly place the responsibility for managing daily life in the hands, as it were, of intelligent machines, we also have put them to work solving previously unsolvable problems. Their ability to extend the human mind and manage complex computations has been integral to the explosion of technology and science in recent decades.

Computers also have been able to extend our senses—providing reading machines and other devices for people with disabilities, giving artists and designers powerful new tools to compose music, design graphics, and produce unbelievable special effects. They have also allowed us to communicate with each other more efficiently and quickly over greater distances than ever before. Robots, guided by computers, have taken over jobs that are either too dangerous or too tedious for humans to endure, and they have already begun to extend our reach into space by exploring worlds where humans cannot yet go.

PARALLEL PROCESSING

Imagine a difficult and complicated task. Then consider, would it be quicker for one person to do each part of the task, one after the other? Or would it make more sense to split the chores among several people who will all work on their part of the task at the same time? Under most circumstances, using more people would get the job done more quickly. This is the principle behind computers that use parallel processing.

The computers that we use at home are like one person doing one part of a task at a time. But in research and business there are tasks whose size and complexity require computers than can do more things faster. These supercomputers contain more than one processing unit which, under the supervision of a central processor, work side by side on separate parts of the task.

Speed and power are the defining characteristics of these supercomputers, and they get the job done in a variety of ways. Single Instruction, Multiple Data computers do several parts of a task by applying one set of instructions to different chunks of data. Multiple Instruction, Multiple Data computers are able to give a separate set of instructions for each processing unit to apply to its stream of information. These computers are further classified as fine- or coarse-grained parallel processors. Fine-grained systems have many small processors, whereas course-grained systems have fewer but more powerful processors.

The ultimate parallel processor is the human brain, in which many trillions of cells called neurons work together to remember and evaluate information. The microscopic cells are themselves parallel processors that seem to be capable of many simultaneous calculations. Although researchers have developed parallel processors that can do thousands of computations at a time, they still cannot rival the complexity of the three-pound device we carry around in our skulls.

The most striking example of parallelism in the brain is human vision. Sensors in the eye send millions of simultaneous messages to the brain along bundles of nerves, carrying information that the brain coordinates and translates into shape, motion, color, and detail. It would probably take the power of a billion home computers making 100 trillion computations per second to duplicate what the eye and brain can do.

Researchers have developed a kind of parallel system, called a neural net, that tries to mimic the society of cells in the brain. The machine neurons are randomly connected and designed to make new connections so that they can, in effect, be trained to do things like recognize speech and print. They learn to recognize shapes by comparing what they see through the eye of a camera with images stored in memory. Researchers have also produced neural nets in which separate groups of "cells," each exposed to a different aspect of an image, must help each other to recognize what the image is. Researchers refer to this as a kind of learning, bringing computers another step closer to human intelligence.

ARTIFICIAL INTELLIGENCE

Despite its remarkable ability to process information in some areas, the human brain is impossibly slow at calculating problems that number-crunching computers can do with ease. Yet the best super-computers fail at seemingly simple functions that we humans take for granted. Researchers working in artificial intelligence (AI) are trying to create machines that combine the speed and power of computers with the common sense and subtle intelligence of the human brain.

So far, no one has invented a machine that can think quite like human beings do. Part of the problem is that scientists don't fully understand how the brain works. They have not been able to agree on useful formulae that might define or quantify intelligence. Those abilities that indicate intelligence—and that AI researchers try to program into computers—include flexibility of approach and the ability to recognize and take advantage of lucky breaks, to rank pieces of information according to their relative importance, and to sort out differences and similarities among factors, making sense of contradictory information. Other hallmarks of intelligence include creativity—the ability to generate new ideas or adapt old ideas to fit new situations—and, of course, humor and self-awareness.

Progress in artificial intelligence has occurred mainly in the development of decision-making programs that represent the knowledge of human experts and organize it according to a series of "if/then" rules (see EXPERT SYSTEMS, page 208). AI programs have also been successful in mimicking such individual skills as speech recognition and vision. Computers can play chess, diagnose diseases, and do other things which, when performed by humans, are considered to be intelligent activities. Nevertheless, it will be a long time before a machine is able to think like a human. For the time being, computers serve best as assistants to humans, not all-knowing replacements.

Advances in microcircuitry and other technologies continue to make computers faster and more powerful, but AI researchers believe that speed is not the key to making intelligent machines. Modern serial computers are capable of processing massive amounts of information very quickly, 10,000 times faster than the brain. But because of the way these computers are designed, they can do only one operation at a time. On the other hand, the average human brain, while slow, is capable of powerful parallel processing.

The brain consists of about 100 billion neurons wired together into a complex network of some 100 trillion connections that can make sense of things in a way that machines have yet to do. For instance, neurons pass information back and forth, storing it in more than one place. That's why memories are evoked by such different sensations as smell, sight, sound, and taste. And it's why memories and abilities aren't always wiped out by a brain injury. In the brain, where neurons die every day anyway, it's standard operating procedure to share knowledge in order to keep it alive. Then when an injury occurs, the surviving neurons take over the work of those that were lost.

Furthermore, neurons seem to have a finely tuned system of checks and balances so that when one neuron goes wrong, others correct for its error. In comparison, computers are deterministic mechanisms. One tiny goof can snowball into a big mess (see CHAOS, page 69). One of the challenges facing the designers of intelligent machines is building in flexibility and self-correction.

Although AI has not produced all of the technological miracles that were first predicted, it has forced researchers to examine more closely how humans see, understand language, and use knowledge to solve problems, resulting in whole new fields of study, such as cognitive psychology. Some very useful devices and programs have already been developed, and researchers have also developed machines that can recognize and respond to human speech.

EXPERT SYSTEMS

One of the most widely used products of AI research is the expert system. Important knowledge and experience in a given field is incorporated into a special program that can solve problems, much like having an expert in your computer. There are expert systems that can diagnose illnesses, solve mechanical problems, and locate oil or mineral deposits. Sometimes they do a better job than human experts, but they are most useful in assisting humans by sorting out and predicting trends in large amounts of data.

To lay the foundation of an expert system, specialists, called knowledge engineers, interview experts to find out what they know. Then they put together a data base, or body of facts, complete with the decision-making rules that go with them. Programmers design a program that poses if/then questions to the data base so that a less experienced person can use the expert system to solve problems.

These programs aren't easy to design because experts rely heavily on intuition, which is difficult to analyze and quantify. Critics of expert systems claim that it's impossible to write computer programs that take into account the subtle gradations of meaning involved in impressions and hunches. However, AI researchers are trying to do just that by converting gradations of meaning to mathematical language. They attach a certainty factor, or number, to each piece of knowledge indicating how certain the expert is about it.

Successful science and medical-expert systems developed in the 1970s and 1980s laid the groundwork for this technology's acceptance in the business world. MYCIN, an expert system that diagnoses infectious diseases, was tested against the diagnoses of doctors at Stanford University, and the results were judged by people who did not know which diagnoses came from a machine. The expert system turned out to be as good or better than the human diagnosticians. But while many medical expert systems have been developed, doctors have been reluctant to use them.

Business, however, has embraced the technology; thousands of systems have been developed and their use is growing. American Express uses an expert system to evaluate credit applications, finding that it reduces costs by saving time and eliminating bad decisions. General Motors and Ford use expert systems to give their mechanics access to the latest information about the increasingly complex automobile. Expert systems have proven useful in designing complicated computer systems that otherwise would have required the work of hundreds of humans. A system called Prospector has proved capable of finding geological deposits that human searchers overlook.

The U.S. Defense Department has shown interest in this technology as a way to manage weapons, reconnaissance aircraft armed with computer vision systems, naval battle planners, intelligent systems that help pilots control their high-tech aircraft, and robot vehicles.

The U.S. Navy began using an expert system in 1987 to coordinate the movements of hundreds of ships and thousands of aircraft over 95 million square miles of ocean. The Force Requirements Expert System has proven to be quicker and more accurate by producing, in minutes, reports that used to take hours for human planners. During 1991's Desert Storm battle in the Middle East, we all saw the impact that intelligent machines can have on warfare.

FUZZY LOGIC

Life is full of uncertainty. When a father, calling from the doorway at dusk, instructs his child to "come in pretty soon," he is giving an order that is less than precise. Does he mean ten minutes, thirty minutes or two hours? The child, having learned what "pretty soon" means, can deal with the imprecise nature of Dad's instructions. He or she had better go inside in about ten minutes.

Fuzzy sets—groups of numbers that do not have clear boundaries—were invented to represent such imprecision in life by converting gradations of meaning into mathematical language. For example, what the child intuitively knows would be converted into a set of times that would satisfy her father's idea of "pretty soon." That set might contain times anywhere from five to fifteen minutes. Fifteen minutes would probably be pushing Dad's patience a bit, while five minutes or less is unrealistic. A fuzzy set would contain times close to ten minutes and reflect the degree to which each time satisfies the requirement of "pretty soon." In this way, shades of meaning can be conveyed in language that a machine would understand.

Research in artificial intelligence has produced computer programs that attempt to accomplish tasks as humans would. But the all or nothing language that computers use—a binary number system in which one means on and zero means off—is not good at expressing the imprecision of human thought and language. Fuzzy logic allows for the range of possibilities between zero and one or yes and no. Sophisticated computer programs called expert systems, in which the imprecise and relative nature of knowledge is converted into fractions of certainty, sometimes make use of fuzzy logic. These programs are difficult to design, because experts rely heavily on intuition, which is difficult to analyze and quantify. We humans don't always understand how and why we know what we know.

Fuzzy logic has proven useful in the design of cameras, computers, air conditioners, refrigerators, and washing machines, mechanisms that need to respond in a more fluid way to changing conditions.

ROBOTICS

For hundreds of years we have been waiting for robots to walk into our lives. But our fantasy robot, a mechanical image of ourselves, has proven to be unsteady on its feet. The humanoid slave-labor force that science fiction and some researchers predicted has not materialized. Instead, there has been a quiet technical revolution. We can stop waiting for the robots because they have already arrived.

The working robot, however, does not look like we expected it to look. The definition of robot has changed from form, a kind of intelligent metal human, to function. For those who take the broad view that a robot is any mechanism that senses, thinks, and acts, robots are everywhere. Appliances, cars, oil refineries, or any system that uses computer technology to function automatically can be called robotic.

Thirty years ago experts predicted that factories would be the first great proving ground for robotics, which would revolutionize industry, displacing workers and increasing productivity. Futurists predicted busy robotic factories running night and day, free of human complications. While that has not happened, the biggest use of robots is in industry, where they exist most often as jointed arms that weld and pump fluids. Japan leads the world in the use of industrial robots, followed by the United States.

Although their use requires expensive initial changes in factories, robots have proven to be helpful for a number of reasons. Operating from stationary platforms or tracks, robots do a more consistent job at a task that is repetitive, boring, and dirty. They don't need ventilation, heat, light, and other safety considerations at their work stations, as people do.

For example, Ford Motor Company uses robots to paint trucks. Its robots are huge orange arms with seven ranges of movement that slide back and forth on tracks as they paint the truck bodies passing by them. A red sensor glows at the end of each arm as it bobs up and down, back and forth, in precisely programmed movements. Behind a glass wall, operators press buttons on control panels, and a computer screen spells out the procedure the robots are following. The robots can change the paint colors or patterns they are using in seconds, and they can work twenty hours a day surrounded by clouds of paint fumes.

Small numbers of robots are already at work cleaning up radioactive and toxic wastes, exploring distant planets, laying cable, and photographing shipwrecks at the ocean's bottom. They are guarding factories, patrolling military bases, and delivering mail. But the next frontier for large-scale robotics may be service industries and agriculture.

Researchers in Wisconsin developed a fast-food robot that can flip burgers and cook fries, while scientists at Carnegie Mellon University in Pittsburgh have developed a robot that makes pizza. In Florida agriculturalists have been experimenting with robotic orange pickers, in Holland and Germany researchers are working on robotic milking machines, and in Australia they have devised a robot that can shear sheep.

A benefit of robotics research has been a better understanding of human functions, and the newest developments in robotic vision and other sensing devises contribute to and depend upon advances in understanding these functions in humans. It's clear that it will be a long time before robots duplicate the abilities of a person. In fact, some researchers think that machines won't have anything quite like human intelligence or consciousness until they experience the world in the same physical way as humans do—moving about, touching, smelling, seeing, and hearing it.

Making a machine that can do these things, however, and all the other complex sensing and decision-making tasks that people can do is incredibly difficult. For example, a small child can recognize the sound of her mother's voice and its message telling her to pick up the toys in her room. The child, if she wants, can easily pack wooden blocks into a box that is kept on her shelf. But she can't reach the block box and, remembering what happened the last time she tried to climb up the shelves to get something, chooses to create the appearance of neatness by stacking the blocks into a tower in the corner. A robot, however endowed with the most sophisticated computer brain, would fall all over itself trying to complete this apparently simple task, which requires speech recognition and processing, pattern recognition, motor skills, imagination, and flexibility of approach.

Some researchers think that the attempt to capture the mechanics of human thought in a humanoid device is misspent energy. They contend that making machines that can get around and work in the real world is more important than trying to make a device that is conscious, in the human sense. Groups of dumb robots could work together to do jobs, reacting and adapting to their environment, while never being fully conscious of it.

Some fantastic roles for robots are still being predicted, however. Futurists and AI experts say that someday we may use lifelike robotic images of ourselves to communicate with others over great distances. Our robotic surrogate could stand in for us physically while we direct its movement by remote control.

The trend in technology to shrink things to microscopic scale extends into the world of robotics. It is already possible to make microscopic machines that actually function, although very simply. Someday, armies of tiny robots could be mobilized in our bodies to repair such problems as scouring cholesterol from our blood vessels or fighting wars with cancer cells. We won't see these robots as images of ourselves. In fact, we won't "see" them at all.

VIRTUAL REALITY

You slip into a special suit, baseball cap, and goggles and suddenly find yourself in a strange new world. When you point a gloved finger, your body seems to fly across the room and open the window onto a purple landscape. You can rise effortlessly through the ceiling and gaze down at the roof of the house and the mountainside where it sits. Your new outfit, which is tethered by cables to a supercomputer, is threaded with sensors and actuators—devices that create pressure, motion, temperature, and other sensations under computer control. This outfit and the complex software it runs is called a reality engine. Your goggles shut out the everyday world and give each eye a screen that is fed images; you become part of a three-dimensional computer graphics world.

As you move, sensors in your suit communicate body movement

to the reality engine, which reacts almost immediately to your change of position in the space. When you turn your head, you see a new aspect of the reality. It is possible to pick things up and move them around. Actuators in the gloves provide the sensation of touching and holding objects. Through earphones you are immersed in the three-dimensional sound of this new reality.

It sounds like science fiction, doesn't it? But it's not a dream. You've entered virtual reality. Like the flight simulators pilots use to practice flying, virtual reality (VR) is a way to experience something without actually doing it. Reality engines exist right now in a number of research facilities around the world, although the technology is still in its infancy. This doesn't prevent researchers from predicting amazing applications for VR.

Most of us exchange information with computers while staring at a screen. Virtual reality allows us to step through the screen and insert our physical selves into the program. In fact, computer screens and keyboards may someday be obsolete. VR technology represents a new frontier in user friendliness. It's meant to dissolve the barriers between computers and their users, allowing humans to interact with information in a more natural way. It may be especially useful as a method of visualizing data. Scientists, business analysts, and others who examine large amounts of data could turn numbers into visual realities like trees or hills in order to see better the patterns that predict trends.

VPL Research in Redwood City, California, a pioneer in VR research, developed and is now marketing the DataGlove, a sensor-laden glove that conveys texture, while at the University of North Carolina at Chapel Hill, chemists are using VR to find new ways to combine atoms and molecules into medicines. The Japanese and the British are working on telerobotics, combining VR technology with robotics. By interfacing with robots, workers could see and maneuver within a dangerous environment like nuclear or toxic waste, fires, or warfare.

NASA's Ames Laboratory near San Francisco, where VR began, hopes to use telerobotics to explore real worlds that are unreachable by or hostile to human astronauts. Instead of using a made-up world that is stored in a computer program, NASA hopes to use robots with camera eyes and sensor-laden bodies to stand in for human explorers. A robot could carry out tasks on the surface of a planet while its human counterpart guided its activity from the safety of a space station or planet Earth. Scientists could accompany and lend their expertise to astronauts on the surface of Mars by way of VR without actually having to go along.

Someday people thousands of miles apart could "meet" in a virtual reality environment and work together on a project, with the technology allowing them to assume any physical identity they chose. For example, two architects, presenting themselves as virtual polar bears, could work together to design a zoo building and test the virtual version of the structure before it's built. Virtual reality's application to the design of such things as buildings and aircraft could also eliminate unforeseen problems that become apparent only to those who must use or maintain them.

Some scientists envision using VR to create new chemicals by manipulating atoms and molecules like Tinkertoys.™ Surgeons could train for intricate procedures by practicing on virtual patients. Meanwhile, psychologists and other researchers are trying to ascertain whether training in virtual reality will transfer successfully to the real world. If it does transfer well, the technology may also be useful as a way for one surgeon to work hand in hand with another, miles away, guiding him or her through a difficult and unfamiliar procedure. Developers of VR technology foresee physicians being able to use VR to travel through blood vessels and make repairs on an ailing heart.

This technology can also be fun, and entertainment industries may make the greatest use of VR. Someday we may be able to buy time in a VR world and embark on such adventures as a day-trip to the moon or some place that exists only in the mind of a creative programmer. Theater may become an interactive experience, while art museums might allow you to walk into the painting or hide inside the head of a statue. Music could become visual and tactile.

In sports centers based on virtual realities, you could play games with virtual opponents, run on a treadmill or ride a stationary bike through magical landscapes, and achieve things you could never do in an ordinary gym. Sports training systems are being developed to improve your golf or tennis game and teach skiing without the benefit of the bunny hill.

Television and movies may cease to be a passive entertainment, as you step right through the screen. The appeal of today's video games is nothing compared to the spell that reality engines may cast on young people. Parents no doubt will fret over the hours their children spend muttering and gesturing while connected to portable reality engines. We should expect controversy along with the fun.

But maybe educators will harness this technology to take students beyond "hands on" and allow them to experience virtual versions of what they are learning. Students might walk beside soldiers in the jungles of Vietnam, visit a space station, or witness the replication of DNA within the nucleus of a cell. The future beckons.

LASER

Ordinary light is a scattered mishmash of different wavelengths bouncing all over the place. Laser—short for Light Amplification by Stimulated Emission of Radiation—stacks waves of light together in one coherent beam, concentrating and optimizing their energy.

Light is created when atoms release energy in the form of photons. When an electron that is orbiting the nucleus of an atom jumps down to a lower orbit, it lets go of its energy by throwing out a photon. Each kind of atom releases energy at a characteristic wavelength. Radiation from the sun, porchlights, and campfires all represent more than one kind of molecule vibrating at different energy levels and throwing out photons that have different wavelengths.

Lasers use specific atoms that are excited by some stimulus—a light, a chemical reaction, a beam of high energy electrons, or another laser beam. When these excited atoms are blasted with a photon that is vibrating at the same energy level, the electrons are forced to jump down to a lower level, at the same time throwing out their excess energy baggage in the form of photons. These released photons strike other atoms, causing the release of still more photons.

The chamber in which all this is happening, usually a glass rod or tube, is fitted with mirrors that are positioned so that the photons will bounce back and forth between them millions of times, causing their wavelengths to become coherent, that is, to fall into step with each other and align themselves into a neat stack. The mirrors are positioned so that some of the photons are allowed to escape as a narrow beam of color.

The first laser was made in 1960 by American physicist Theodore H. Maiman. Since then, these devices have become important tools in science, industry, and medicine. In our homes, lasers are most likely found in the stereo cabinet, where they play compact discs. CDs are manufactured by converting energy waves into laser pulses, which etch invisible craters into metal discs. The craters spell out the pattern for sound (and pictures, too, in laser video discs) in a binary code (ones and zeros) that minutely plot waveforms so that they can be reproduced with extreme clarity. Later, the craters, or dots on the disc, are read by a low-energy laser in the CD player and converted back into energy waves or music, speech, and pictures.

Another everyday application of laser beams is the bar-code reader at the supermarket. A very low energy beam reads the black price code on grocery items and transmits it to the computerized cash register. Bar codes are being used increasingly to identify and manage materials. For example, your local library may be using them to keep track of books. Laser beams are also used in communications to send thousands of wave-encoded messages through fiber-optic cables, which are cheaper and more compact than copper-wire cables. (See FIBER OPTICS, page 219.)

A still experimental type of laser may someday be used in telecommunications and computers. Scientists have developed micro-

Waves that are same length but are not coherent or stacked together.

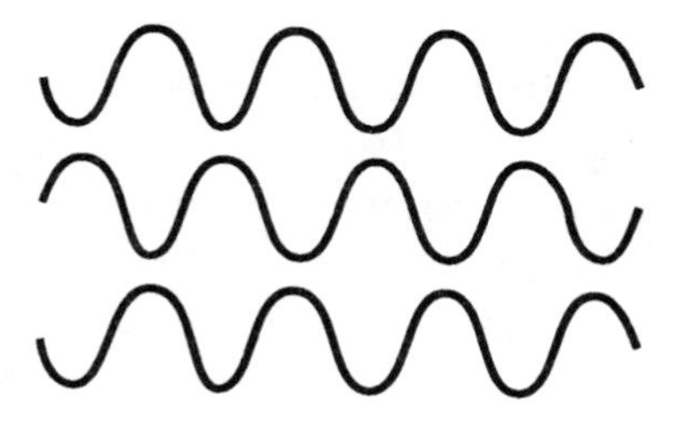

Coherent Waves

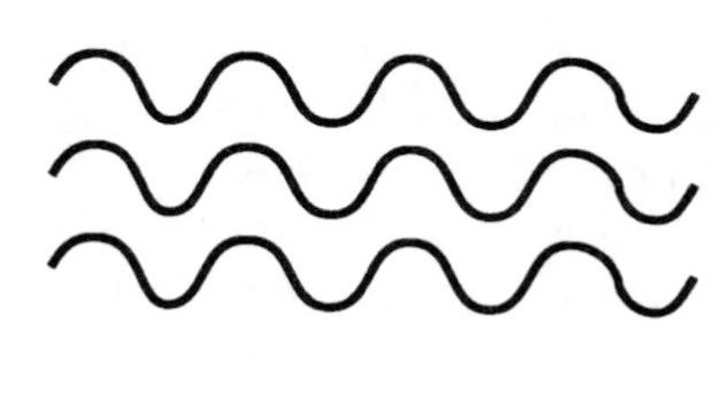

scopic lasers, 10,000 of which would fit on the head of a pin. Thumbtack-shaped, club-shaped and columnar microlasers are carved out of ultrathin layers of semiconductors. Millions of them can fit on one chip and may become part of a computer that operates with pulses of light, rather than electronic currents and switches. Experts say that such an optic computer would be faster and more powerful than any of today's supercomputers.

Lasers are being used in fusion research bent on finding new sources of energy. Laser fusion reactors attempt to force fusion by means of temperature and pressure by blasting a pellet of frozen hydrogen with ten or more laser beams. The lasers are arranged so that the force that strikes the pellet compresses the material symmetrically, raising its temperature to many millions of degrees. The goal is to trap the hydrogen atoms in a situation in which they have no choice but to fuse into helium atoms, releasing energy in the process. (See FUSION, page 60.)

Printers that use laser technology are fast and quiet, and they produce high-quality print. In this case, the laser is used to transfer a page of print onto a charged surface, removing the charge in such a way that a negatively charged pattern of the printed page remains. This surface is exposed to thermoplastic particles that are attracted to the charged pattern. Then a positively charged piece of paper passes over the charged pattern, attracting the particles that are than melted into place by hot rollers.

Chemists are developing a way to use lasers to break apart and rearrange chemical bonds within molecules, giving them greater control over chemical reactions. And researchers continue to develop many medical applications for lasers. For example, these devices can vaporize cells and coagulate proteins in such a precise way that they become suitable for delicate surgical procedures.

Opthalmologists use lasers to repair torn retinas and to stop blood-vessel leakage in an eye compromised by diabetes. The pressure from fluid buildup in the eye, a symptom of glaucoma, can be relieved by using lasers to bore tiny holes for drainage or eliminate blood vessels that are blocking the natural system.

By threading optical fibers into the body, surgeons are able to direct lasers at tumor cells and destroy them. Researchers are developing techniques to enable physicians to clear arteries of plaque with fiber-optic lasers in combination with balloon angioplasty, a procedure in which a catheter tipped with an inflatable balloon is threaded through a clogged cardiac artery; the laser assists in clearing the blockage and the balloon is inflated to compress the plaque against the artery walls. Fiber-optic lasers are used for brain and head surgeries, in which greater precision is also needed, to reduce swelling and bleeding, both of which can cause brain damage.

High-power lasers are used in industry to carve and weld, cutting through wood, many layers of fabric, or thick plates of steel as though they were butter. Laser beams can also be used to measure vast distances through space. And finally, a mirror placed on the moon during a 1969 moon walk still bounces laser beams back to Earth, giving scientists a very precise measurement of the distance in between.

FIBER OPTICS

You've probably seen children's flashlights and decorative tree lights in which plastic fibers spill colored light from the ends of flexible filaments. Other materials, including human hair, can carry light in this way. But threads of glass, as fine as hair, can carry light uninterrupted for hundreds of miles under the ocean or along several circuitous feet inside the human body. It's these fields of communications and medicine that reap the most benefit from fiber-optic technology.

Optical fibers are made of extremely pure silica glass, with a special coating or cladding to prevent captive light from straying. They can carry hundreds of telephone messages on bundles of fibers a fraction of the size of traditional copper-wire cables. The messages are translated from electrical information into pulses of laser light, which are carried by the fibers and translated back into electrical information at their destination. Fiber-optic telephone links connect many major cities and cross under the Atlantic Ocean to Europe.

By means of fiber optics, endoscopy—derived from Greek roots meaning "observation within"—peers inside the body to examine organ systems and perform certain therapies. Fiber-optic endoscopes are flexible devices, no wider than a pencil, that contain two bundles of optical fibers—one to carry high-intensity light into the body, and the other fitted with a lense to transmit images back out again. Physicians are able to see high-resolution color images of the lungs, heart chambers, intestines, bladder, and stomach by threading the endoscope into a natural orifice or through a small incision. Each of thousands of fibers in the imaging bundle "sees" a little piece of the image, carrying it out to an eyepiece, camera, or screen. The procedure often eliminates the need for a physician to cut into a patient to find out what's going on inside. Researchers say this is especially useful with children, the elderly, and patients for whom surgery might be too risky.

Extra channels in endoscopes allow physicians to inject water or air for cleaning away debris in order to get a better view and also for removing samples of body fluids for testing. Extra channels in the endoscope can be used to thread tiny surgical instruments into the body to nip off and grab tissue samples. Such an instrument is used to remove colon polyps, which are linked to cancer.

Laser beams can be sent into the body on fibers to seal off or cauterize ruptured blood vessels and stop internal bleeding, heating proteins in the body tissues so that they solidify. Fiber-optic lasers are also used to clear away blood clots and plaque—the fibrous, fatty deposits that clog arteries. Physicians are able to choose from various types of laser beams in order to target specific tissue, whose pigmentation determines the energy waves that are absorbed.

A tissue's tendency to absorb energy because of its color is the basis for other fiber-optic laser therapies. Fluorescent endoscopy is a technique used to find small and otherwise undetectable tumors. The patient is injected with a chemical dye that is absorbed by tumor cells, causing them to fluoresce, or glow, when exposed to ultraviolet energy waves piped into the area on quartz fibers. Once the tumor cells are found, they are treated with a kind of laser energy that excites the atoms and molecules in those cells suffiently to kill them.

Fibroscopes are used as sensing devices that can be threaded into the bloodstream, where the way in which light is reflected provides important information about body chemistry. For example, it is possible to determine how fast blood is flowing by bouncing light off moving blood cells. A computer compares the wavelengths of light that are sent into the blood vessel with the wavelengths of light reflected by the cells, which will be different. Fiber-optic sensors are also used to measure pH and pressure in other conduits and organs in the body, as well as the presence of microbes, oxygen, hormones, toxins, enzymes, and antibodies.

Finally, fiber optics play a role in the so-called Band-Aid™ surgeries, in which several small incisions are made to accommodate not only the fiber bundles but also the instruments used to perform surgery. Laparoscopy, as it is called, is used for a number of abdominal surgeries, including hysterectomy and gallbladder removal. Laparoscopy's supporters say it shortens recovery time because it eliminates large incisions and reduces the need for blood transfusions.

BAR CODES

When you buy groceries at the supermarket or check out a book at the library, chances are you're observing Optical Character Recognition (OCR) at work. These computerized devices "read" information by scanning printed characters or groups of vertical lines called bar codes with low-powered lasers (See LASERS, page 216.) Most Americans remember taking basic skills tests in school and having to use a No. 2 pencil to darken oval shapes on their answer sheet. These tests were read optically by machines.

The dark and light areas on the answer sheet and other material read by OCR devices reflect the laser beam back to the scanner. The scanner processes the different ways in which the light is reflected and sends the information to the computer in the form of electronic pulses. The computer compares this information to data already stored in memory and, in the case of the test, notes correct answers and tallies the score.

At a grocery store, the computer tells the cash register to enter and display the product's name and price. This technology saves time and reduces errors at the check-out counter, but it also helps the store in other ways. The computer deducts the sold item from the store's inventory and alerts the stock manager when it's time to replenish supplies of that item. Other facts about the product, such as special sale prices or quantity discounts, can be programmed into the computer and made available to the clerk at the cash register.

In bar codes—also called universal product codes or UPCs—the thick and thin black lines, as well as the white spaces between them, represent binary numbers (0 and 1, standing for on and off) that are combined in a code and translated into letters and numbers. Special lines are used to indicate the beginning and end of the code, so that no matter which way the object is passed over the scanner, the computer will know the proper order in which to interpret the symbols.

Bar codes are printed on most consumer goods and are increasingly used where it's important to identify and keep track of individual items. For example, libraries are able to use computers to manage circulation by giving each book and each library card a bar code. OCR is also used to scan and store literary works and other printed material on computer disks and it's used by post offices to electronically sort mail. Banks use magnetic ink to print codes and dollar amounts on checks which are then read by special OCR machines.

This technology has been used to develop machines that can read to the blind. The Kurzweil Reading Machine scans printed material and sends it to a computer. Special software recognizes the letters, puts them together into words and tells a speech synthesizer (an electronic device that mimics human speech) how to pronounce them.

ULTRASOUND

An ultrasound machine works rather like a bat that is trying to find its way in a dark and unfamiliar cave. It emits high-frequency sounds, which bounce off objects and provide an echo map of the space. While a bat's squeaks are emitted at frequencies of 100,000 hertz (Hz), or cycles per second, ultrasound (which means "beyond sound") is actually anything above 20,000 Hz, the limit of human hearing.

These high-frequency sounds are useful in a number of ways, especially in medicine, where they are used in noninvasive medical imaging devices. The ultrasound machine takes "pictures" of babies in the womb, detects gallstones, evaluates blood vessels, and guides needles into organs to take biopsies of tissues.

In this technique, a high-frequency electrical current is passed through a piezoelectric crystal. (*Piezo* is a Greek prefix meaning "to press or squeeze.") Certain crystals, like quartz, will produce electricity when placed under pressure and can convert mechanical energy into electrical energy and vice versa. The crystal vibrates and converts the electrical energy into sound waves, which are directed into the body where they bounce off the varying densities of this internal landscape and return to the machine. There the echoes are translated into an image on a screen.

Narrowly focused ultrasound can be used to destroy such solid masses as kidney stones, without surgery. In a technique called extracorporal shortwave lithotripsy, or ESWL, the soundwaves are set to resonate at a specific number of cycles per second, causing certain molecules to resonate while leaving others alone. The vibrations pulverize the unwanted material. Sound vibrations also can be used clean small, delicate objects, as well as surgical and dental instruments.

MAGNETIC RESONANCE IMAGING (MRI)

Magnetic Resonance Imaging (MRI) is a diagnostic imaging technique that rearranges the hydrogen molecules in living tissue as though they were iron filings on a sheet of paper. By lining up these atoms, which are as plentiful as water, and bouncing radio waves off them, MRI technicians are able to create three-dimensional images of the internal workings of the human body without any known major risk to the patient. Most other imaging techniques, except for ultrasound, use radiation, and thus involve some risk to the living things they are designed to help.

MRI takes advantage of the fact that human bodies and other living tissue are made primarily of water. In every water molecule there are two hydrogen atoms whose nuclei, when exposed to a magnetic field, will orient themselves with the lines of the magnetic field. These hydrogen atoms, like all other atoms, resonate or vibrate at a frequency particular to them. While exposed to the forces of a very powerful magnet, the tissues are also bathed in radio waves of the same frequency, so that the hydrogen atoms' nuclei rearrange themselves, turning away from the magnetic lines. But when the radio waves are turned off, the hydrogen atoms bounce back into place along the lines of the magnetic field. At the same time, the hydrogen atoms emit tiny electrical signals that are collected by a receiver, which measures the speed and numbers of atoms that respond. This information can then be analyzed by a computer.

But how does all this create a picture that doctors can use to diagnose illness or better understand the behavior of the cell? The answer lies with water, which exists in the body in varying concentrations. There's more in blood and urine, for example, than there is in bone or cartilage. The percentage of water in different organs and glands also varies. So the corresponding differences in concentration of hydrogen atoms are reflected in the number of signals emitted when the radio waves are taken away. By examining an area of the body from several angles, a computer can build a three-dimensional picture of the tissues inside the body.

Hundreds of MRI machines are in use in the United States and are especially useful for obtaining early diagnoses of cancer and circulatory diseases, and for assessing the effects of drugs on the body. The strong magnetic forces are created by superconducting magnets that are cooled constantly by liquid helium. Certain conductors, when cooled down to nearly absolute zero, are capable of passing current without any resistance at all. If the temperature is maintained, the magnet need only be powered up once and should operate literally forever. The newest ultrafast magnetic resonance imagers can gather many images in a second, making for a sharper image because it eliminates the blurring caused by body movement.

Another recent modification of MRI makes three-dimensional

movies of the heart as it beats, allowing researchers to see how the muscle functions and how it behaves when it doesn't function properly. A team of researchers from different disciplines—radiology, cardiology, physics, and engineering—created a device that alters the orientation of selected sections of hydrogen nuclei in the heart. These sections appear as stripes on the computer image. As the heart beats, the stripes distort, allowing the observer to see how the muscle twists and untwists. Researchers will be able to see how the behavior of the heart muscle changes when it is diseased or injured.

The most powerful magnetic resonance imagers are up to 80,000 times stronger than Earth's magnetic field, enabling researchers to monitor the minute chemical changes that occur inside cells. They are so powerful that they have been known to tear metal buttons off of clothing. MRI is considered generally safe, although some scientists are concerned about the effect that changing magnetic fields might have on the nervous system, especially in patients with heart disease or a seizure disorder.

COMPUTERIZED AXIAL TOMOGRAPHY (CAT SCAN)

Like opening a window into the body, Computerized Axial Tomography (CAT or CT) is a variation on X-ray imagery. The machine spins a camera around its subject, scanning with a flat, thin beam of weak X rays, taking pictures from many perspectives. A computer shows thin cross sections of the subject or puts them all together to build a three-dimensional picture on a monitor screen.

The name of this diagnostic technique, like most scientific terms, has Greek and Latin roots. *Tomo* means "cut" and *graph* means "to write." Axial comes from the Latin word for axis or axle. A person having a CAT scan must lie down on a table that is designed to move slowly through a tube or doughnut-shaped device containing an X-ray scanner and a sensor that collects information and communicates it to the computer. In short, the scanner spins around an axis and produces computer-graphic images of thin cross sections of the body.

A CAT scan is more useful than a regular X ray because it gets around the denser objects, like bones and muscles, which can block the view, and it sidesteps the chemicals in the body that tend to absorb X rays and blur or obscure the image. The high-resolution digital image can detect tissue abnormalities as small as a millimeter. Because the subject is surrounded, the CAT scan also depicts the dimensionality of an organ or tumor, information that may be very important in diagnosis and treatment. And, it does this task with much lower doses of radiation than are required to expose X-ray film.

CAT scans have been used to study the shape and interior surfaces of the skulls of early hominids (ancestors of ancient humans) and dinosaur fossils by seeing through the dirt and rock clogging the fragile remains. Researchers have also been able to use such information to reconstruct the missing portions of such finds.

More powerful scanners are used in industry to inspect structures and machine parts to find otherwise invisible flaws. A high-speed CAT scanner is being developed to screen baggage for airlines: It would be able to take cross-sectional images of objects like books and tape players, comparing dimensions, density, and other characteristics with a data base of known explosives, thus reducing the danger of terrorist bombings.

POSITRON EMISSION TOMOGRAPHY (PET)

By injecting radioactive substances into the bloodstream, researchers are now able to see blood flow and watch the body's metabolism in operation by means of Positron Emission Tomography (PET). Like the CAT scan, PET shows cross sections of the body, but unlike the CAT scan, it does this without exposing the body to radiation from the outside. Instead, the PET scan does its work by detecting emissions from very low-level radioactive materials injected *into* the body.

A positron is one of a family of subatomic particles called antimatter. Physicists have discovered that every particle of matter has a corresponding particle that is a twin in every way except for its electric charge. When they meet, these positive and negative particles will cancel each other out, leaving only a burst of energy as the by-product. So when a positron meets up with its opposite, the electron, which has a negative charge, they annihilate each other in a flash of gamma rays—high-energy waves, similar to X rays, that are emitted during interactions between nuclear particles.

PET uses the explosive relationship between these two particles to give doctors detailed images of body functions, without having to use such invasive methods as drawing blood or obtaining tissue samples. The patient is injected with a drug or some naturally occurring biological substance that has been tagged with positron-emitting radioactive atoms. The atoms, which decay or radiate energy very quickly, are created in a nearby cyclotron or particle accelerator immediately before the test. (See PARTICLE ACCELERATOR, page 49.)

The scanning machine is a ring-shaped device fitted with gamma ray–detecting crystals. The patient lies inside the ring, and when the injected radioactive atoms begin to decay or shed energy, positrons collide with nearby electrons in the body and the flash of gamma rays is noted by the detectors, which light up. A computer collects the detectors' flashes and converts them into a colorful graphic representation on the screen. The observer can see how a substance, radioactively tagged in this way, is absorbed by the body, or how blood flows through certain vessels in the brain.

PET has been especially useful in diagnosing problems and understanding activity within the brain. In particular, physicians use PET to evaluate malignant tumors, stroke damage, and the effects of brain disorders like Alzheimer's and Parkinson's disease, as well such disorders as schizophrenia and epilepsy.

PET was invented in the early 1970s, by biophysicists Edward Hoffman and Michael Phelps at the University of California at Los Angeles School of Medicine during which time it was used by physicians primarily as a diagnostic tool. More recently, researchers have turned to this technology to learn more about how the brain functions. Some studies have attempted to map the regions of the brain where specific thought processes take place by measuring increases in blood flow and the consumption of oxygen and glucose, the fuels for working brain cells. The more these cells work, the more fuel they use, and this is visible in a PET scan.

Using PET, researchers have attempted to trace the sequence of activities in the brain that seem to process language, and to map the areas of the brain that get involved in processing information. Critics of the technique say that blood flow changes may not be a clear indicator of neural activity, which may be much quicker and more subtle.

Despite the criticism, researchers are now able to use PET to study processes in the living human brain without risk to its owner, and learn things about its function that previously could be inferred only from animal studies.

DIGITAL SUBTRACTION ANGIOGRAPHY

In an angiogram, catheters are used to deliver an opaque dye (one that doesn't allow X rays to pass through it) to a spot in an artery where a narrowing or some other problem is suspected. The dye shows up on an X ray and reveals the dimensions of the blood vessel. This test can provide physicians with valuable information about changes in blood flow to a diseased organ or tumor. A variation on this technique, called digital subtraction angiography, uses computerized images to show the flow of blood through vessels.

First, a live X-ray image, or movie of X-ray pictures, is made of the area. This image is digitalized—that is, broken into tiny elements, like the dots that make up a newspaper picture. Hundreds of thousands of dots, each representing a quantity of brightness, are stored in a computer. Then dye is fed into a vein by way of a catheter and another X ray is taken. It, too, is broken into dots and stored in the computer, where it is compared dot by dot with the first image. When the dots match, they cancel each other out as white space. The range of difference between the dots in the two images becomes increasingly gray until great differences appear to be black. In effect, the differences of one image are subtracted from the other, leaving only a digital computer image of the dyed blood flowing through vessels. All unwanted information is left out. It's then possible for doctors to see how blood is flowing through large as well as quite small vessels.

Digital subtraction angiography is useful in detecting restrictions on blood flow to the brain and kidneys. It can also evaluate the flow of blood to the heart muscle and the extent of disease, as well as detect the threat of heart attack.

This procedure is considered safer than other techniques in which catheters are inserted farther into blood vessels, with more risk of injuring walls. It is relatively quick, performed with a local anesthetic on an out-patient basis. The dye causes a brief sensation of warmth in the area where its injected, and some people experience nausea, an allergic reaction to the dye. As we'll see, this technique is very helpful to physicians doing angioplasty.

BALLOON ANGIOPLASTY

Instead of doing more invasive surgery, doctors sometimes are able to open clogged or narrowed blood vessels with a technique called percutaneous balloon angioplasty. (Percutaneous means "through the skin," and angioplasty means "blood-vessel repair." In the procedure, which is guided by X rays, a catheter or narrow tube is threaded into a small cut in the skin, usually in the thigh, and through an artery to a narrowing or clog in the blood vessel. At its tip, the catheter has a tiny sausage-shaped balloon, which is inflated once the tip has penetrated the partially blocked area. The pressure of the inflating balloon presses the narrowed walls of the vessel to open it up. Or it pushes against plaque, a fatty deposit that can block coronary arteries, flattening it against the vessel walls and restoring blood flow. Balloon angioplasty is used in combination with angiograms, tests that pinpoint the obstruction that needs to be moved.

Although totally blocked coronary arteries are still most often treated through coronary bypass surgery, researchers have developed a technique that threads a laser into the artery to use laser beams to punch a hole in the clog through which the balloon-tipped catheter is then threaded. The balloon is inflated, opening the passageway. There is a risk of damaging the artery walls with the laser beam, however.

Bypass surgery is a riskier procedure than balloon angioplasty and requires a much longer recovery. There is a risk of heart attack associated with interrupting the blood supply to the heart during balloon angioplasty, however, and sometimes it must be repeated because the blockage returns. In about 40 percent of patients some clogging recurs after six months.

Refinements on the catheter have allowed physicians to use balloon angioplasty in small arteries in the arm and leg, eliminating the restricted blood flow that leads to amputation because of the death of tissue the vessel normally supplies. Balloon angioplasty has also been used to open up narrowed heart valves in children. In the latter procedure, called balloon valvuloplasty, a balloon-tipped catheter is threaded into the valve that was deformed by a congenital defect; the balloon is inflated several times to open the valve.

Variations of balloon angioplasty have been used to stop blood loss by pressing against bleeding vessels in the esophagus or stomach, a problem that occurs sometimes in people suffering from liver disease. And it's used to stop bleeding temporarily while the patient is prepared for surgery.

In some cases the balloon is purposefully left inside the body. Filled with a material that hardens quickly, the balloon is positioned and left to shut off blood flow permanently to certain areas. For example, men suffering from infertility caused by varicose veins in the scrotum have been helped with this procedure. The balloon is inserted on the tip of a catheter that is threaded through a leg vein and up into the scrotum, where the balloon is inflated and left to block the internal spermatic vein. When the blood supply to the varicose veins is stopped, they shrivel up. This same technique can block the blood supply to tumors or inject medication directly into the organ that needs it.

SCANNING TUNNELING MICROSCOPE

Engineers from IBM weren't just goofing around when in 1990 they used a scanning tunneling microscope (STM) to spell out I-B-M and draw a stick figure with single atoms. With the widely published photograph of these images, they wanted to show that technology now allows them to do the unimaginable, the sort of thing that previously only nature could do—move atoms around like marbles.

The machine they used was invented in the early 1980s by two scientists, Gerd Binnig and Heinrich Rohrer, who worked for an IBM research lab in Zurich, Switzerland, and won the Nobel Prize in 1986 for developing STM. Their technique allows researchers to make photographs of individual atoms. The microscope is fitted with an extremely fine tungsten needle that scans a few atoms' widths away from the material it's examining. A low electrical current is passed through the needle, causing electrons to "tunnel" back and forth between the two objects. A feedback mechanism notes changes in the gap and maintains a constant distance between the sample and the needle.

As the microscope scans the surface of the sample, the bumps, curves, and geometric arrangements of atoms cause the needle to rise and fall like a record needle in the grooves of an old LP. The texture of the atomic surface of the sample and the subsequent changes in the current passing between it and the needle are registered by a computer, which converts the information into a picture of the atomic dimensions of the sample. With STM it is also possible to see the bonds that hold atoms together.

Later researchers took the technology a step further and used it like a magnet to move atoms around. They learned that by increasing the voltage on the needle they could pull an atom from the surface of the sample and then drop it by cutting back the voltage.

For the time being, this extraordinary technology is not very practical. It requires a massive, expensive machine and superchilled surfaces held in a vacuum. Perhaps a smaller and less expensive probe will be developed to take the place of STM in the manipulation of atoms. If so, the practical benefits may come by the end of the decade.

Meanwhile, scientists are learning how to use STM to build new molecules, atom by atom. Visionaries say it will be possible to make microscopic structures and machines from atoms and molecules, using STM to fasten them together. Extremely small, faster and more efficient electronic components may one day be made by manipulating the conductive properties of materials and also detecting flaws that inhibit the flow of energy. The most likely use will probably be atom-scale information storage, whereby a single 12-inch silicon disc could store the entire contents of the Library of Congress, something that would require 250,000 such discs today.

NANOTECHNOLOGY

Human-made machines are big and crude compared to the natural engines that propel life—DNA, RNA, and enzymes. Nanotechnology is science's attempt to move in on this tiny world and juggle atoms and molecules around with the same precision. The prefix *nano*, which comes from the Greek word for dwarf, refers to the nanometer—a billionth of a meter. (A meter is a little bigger than a yard, or about 40 inches. The average atom is a few tenths of a nanometer in diameter.)

The term nanotechnology was created by futurist K. Eric Drexler, who predicts that someday armies of microscopic robots, given the appropriate feedstocks, could build substances from the atom up, similar to the way in which RNA assembles molecules inside the cell. We would no longer have to send away to other countries for materials and products. We could make them ourselves.

Nanorobots made of proteins could be sent into the body to repair the molecular mistakes that cause illness and disease. They could be used to scrub pollutants from the environment, rebuild livers and arms, and perform microsurgery on delicate tissues like corneas. These tiny workers, Drexler suggests, will also be able to replicate themselves, creating more and more little robots. But if this were true, other scientists say, then horrible weapons could be created using armies of these self-replicating microscopic robots that would devour everything in their path, and could pound the biosphere to dust.

In reality, this degree of sophistication is not likely for a long time. Although researchers have produced very simple microscopic machines the width of a human hair with gears no bigger than blood cells, these micromachines don't work very well, indicating that the concept still needs a lot of work.

But researchers, using Scanning Tunneling Microscopes have been able to move atoms around one at a time (see page 231). Using these very large machines and lasers, biochemists are able to study the atomic behavior of matter and manipulate little bits of it. Low-powered laser beams can be used as tweezers to pick up things as small as the organelles inside a cell and move them. More than just an exercise in dexterity, this work will help scientists understand how cells work, knowledge critical to understanding how cells fight or succumb to disease.

STM and other devices that manipulate atoms also allow researchers to construct molecules one at a time in order to make entirely new materials or improve substances that are already widely used. Ceramics, for example, are heat-resistant but brittle materials that would be more useful if they were flexible. Researchers are combining the properties of ceramics with polymers, large molecules that are flexible and strong. The resulting material could be made into flexible superconducting wires or molded into lightweight parts that would withstand high temperatures like those found in jet engines.

Metals can be rearranged at the atomic level to alter their characteristics. By weaving other atoms into the lattice structure of metal alloys and lacing them with ceramic fibers, researchers are able to make metals that are stronger and more capable of withstanding extremely high temperatures. New composites are also made by layering together thin films of metal and ceramic to make superconductors and other products. Layers, sometimes a single atom thick, are sprayed on top of each other.

What's more, metals and ceramics can be endowed with a kind of chemical intelligence by embedding them with sensors and other materials that detect and adapt to changes in the environment. Materials used in aircraft, bridges, and many other structures could include subtle atomic changes to preclude failure. These "smart" materials might even be able to warn people about structural failures before they happen. (See BIOMIMETICS, right.)

The ability to sense environmental conditions and adjust to them is found in nature, of course, where evolution has produced the strongest and most resilient ways to arrange and glue together molecules. Scientists are looking to biological barriers like skins and shells for recipes and patterns. By copying the molecular structure of certain sea-animal and insect shells, researchers may someday produce sensitive and adaptable skins for aircraft and submarines.

Microscopic patterns in nature are also being used as templates for nanotechnology. For example, a bacterium has a grid of holes, which the organism uses to move substances in and out. Researchers have been able to attach this wall to a graphite base and coat it with a fine layer of metal to make a microscopic screen. The holes in the screen are filled with other atoms to make different kinds of sensors.

Green plants are efficient collectors and transformers of solar energy both of which processes scientists are trying to copy. In photosynthesis, solar energy excites electrons in the chlorophyll molecule, setting off a chain reaction that produces the energy the plant needs in order to produce sugars and carbohydrates. Perhaps someday, similar human-made molecules could be used to solar-power nanoelectronics.

The technical revolution that gave us transistors (which control or amplify the passage of electrical currents) and integrated circuits has set the theme for the electronics of the future: Things continue to get smaller and faster. Today's smallest transistors are already tiny—a hundred times smaller than the width of the human hair—and researchers are working on devices that are thousands of times smaller. By composing crystals, an atom at a time, scientists are working to create extremely small and flawless chips that will continue the miniaturization of electronics. As a result, we may someday be able to carry powerful computers in our pockets.

BIOMIMETICS: SMART MATERIALS AND INTELLIGENT STRUCTURES

Imagine a heavily traveled bridge that adjusts the rigidity of its supports to accommodate the load and vibration of increased traffic. When a support is about to fail, sensors in the bridge alert engineers, who close it down in time to avoid disaster. Nothing quite like this exists now, but someday it might, thanks to biomimetics—research with materials that mime, or copy, biology, intent on giving inanimate objects the sense to adapt under pressure.

Smart materials, as they are called, are designed with some of nature's great successes in mind. They combine components that contribute structure or framework like a skeleton, sensitivity and monitoring like the nervous system, and response like the muscles and other mechanisms in a living creature.

Researchers developing these materials predict airplane wings that change shape in order to achieve lift, rather than relying on hydraulics to move flaps mechanically. Such airplanes could be lighter, faster, more maneuverable, as well as more fuel efficient. Some other innovations that materials research may yield are automobiles with suspension systems that reconfigure their structure to adjust to road conditions; helicopter blades that stiffen in response to turbulence; and submarines that deftly change the characteristics of their skin to reduce water turbulence and to avoid reflecting the sound waves used by enemy sonar. The military is particularly interested in smart materials research because these substances would make aircraft and submarines virtually undetectable by enemy sensing devices.

Such things seem impossible, yet researchers are well on the way to making materials that are embedded with fiber-optic networks that monitor temperature and stress. They are developing electrorheological fluids (*rheo* is a Greek root meaning flow or current.), substances that change from a fluid to a solid state under the influence of electric currents. Shape-memory metal alloys return to a previous shape when heated to a certain temperature, acting like muscle fibers inside structures. Metallic substances called magnetoelastics change in length and flexibility while exposed to certain magnetic fields. They could respond in fractions of a second and quickly bend an airplane wing to the shape that conditions and the pilot dictate.

Smart materials are just part of a revolution in materials research, in which scientists are taking a closer look at the way nature builds things. The explosion of developments in this multidisciplinary field is a result of such technical advances as the scanning tunneling microscope, which allows researchers to examine and manipulate substances at the atomic level. With improved microscopy, scientists are better able to see the components of natural substances and learn what makes them work so well.

For example, researchers are trying to mimic the strength and resilience of bones, teeth, and shell by copying nature's recipe, combining minerals and organic polymers in various brick-and-mortar designs. In nature, the size and arrangement of minerals in their organic mortar determines the characteristics of the composite. Bone is a combination of plate-shaped calcium phosphate crystals glued together by collagen, a protein substance that is also found in connective tissue and cartilage. Tooth enamel, on the other hand, is made of needle-shaped crystals in a matrix called enamelin. Scientists are trying to figure out how the complex structure of such crystals and polymers and their interactions on the molecular level lead to resilient materials like sea shells, teeth and bones.

Abalone, the resilient shell of a mollusk, has been used as a model for composites being designed to overcome some of the structural weaknesses of ceramics—an otherwise useful substance that is heat resistant and tough, yet brittle and difficult to shape. New organo-ceramics combine ceramics and polymers into tough materials that are being tested as tank armor and bone replacements.

The skins of sharks and the exoskeletons of insects are also yielding information that will someday find its way into human-made devices. Insect skeletons are strong shells that are sensitive to temperature, pressure, and other stimuli; they can breathe, and insulate. Someday airplanes may fly equipped with tough, lightweight skins that can monitor and adapt to conditions just as insects do. Already the Japanese have developed an artificial muscle fiber that contracts and expands much like real muscle fibers do.

Ultimately, some researchers predict, building materials may be endowed with the ability to evolve and grow in response to environmental pressures. The possibilities are as endless as the human imagination.

BIBLIOGRAPHY

Allègre, Claude. 1988. *The Behavior of the Earth: Continental and Seafloor Mobility*. Translated by Deborah Kurmes Van Dam. Cambridge, Mass.: Harvard University Press.

Arp, Halton. 1987. *Quasars, Redshifts and Controversies*. Berkeley, Calif.: Interstellar Media.

Asimov, Isaac. 1984. *New Guide to Science*. New York: Basic Books, Inc.

Asimov, Isaac. 1986. *The Dangers of Intelligence and Other Essays*. Boston: Houghton Mifflin Company.

Asimov, Isaac, and Frederik Pohl. 1991. *Our Angry Earth*. New York: Tom Doherty Associates, Inc.

———. 1991. *Atom: Journey Across the Subatomic Cosmos*. New York: Dutton.

Bakker, Robert T. 1986. *The Dinosaur Heresies*. New York: William Morrow and Company, Inc.

Bass, Thomas A. 1990. *Camping with the Prince*. Boston: Houghton Mifflin Company.

Berra, Tim M. 1990. *Evolution and the Myth of Creationism*. Stanford, Calif.: Stanford University Press.

Bishop, Jerry E. and Michael Waldholz. 1990. *Genome*. New York: Simon & Schuster.

Bova, Ben, and Byron Preiss, eds. 1990. *First Contact: The Search for Extraterrestrial Intelligence*. New York: Penguin Books.

Brenner, David J. 1989. *Radon: Risk and Remedy*. New York: W. H. Freeman and Company.

Close, Frank. 1991. *Too Hot to Handle: The Race for Cold Fusion*. Princeton, N.J.: Princeton University Press.

Cohen, Bernard L. 1987. *Radon: A Homeowner's Guide to Detection and Control*. Mount Vernon, N.Y.: Consumers Union.

Cornell, James, ed. 1989. *Bubbles, Voids and Bumps in Time: The New Cosmology*. Cambridge, Mass.: Cambridge University Press.

Davies, Paul. 1983. *God and the New Physics*. New York: Simon & Schuster.

Edey, Maitland A., and Donald C. Johanson. 1989. *Blueprints: Solving the Mystery of Evolution*. Boston: Little, Brown and Company.

Edgerton, Lynne T. 1991. *The Rising Tide: Global Warming and World Sea Levels*. Washington, D.C.: Island Press.

Ehrlich, Paul R. 1986. *Machinery of Nature: The Living World Around Us and How It Works*. New York: Simon & Schuster.

Ehrlich, Paul R., and Ann H. Ehrlich. 1991. *Healing the Planet: Strategies for Resolving the Environmental Crisis*. Reading, Mass.: Addison-Wesley Publishing Company, Inc.

Eldridge, Niles. 1985. *Time Frames: The Rethinking of Darwinian Evolution and the Theory of Punctuated Equilibria*. New York: Simon & Schuster.

———. 1991. *The Miner's Canary: Unraveling the Mysteries of Extinction*. New York: Prentice Hall Press.

Ferris, Timothy, 1988. *Coming of Age in the Milky Way*. New York: William Morrow and Company, Inc.

———. 1992. *The Mind's Sky: Human Intelligence in a Cosmic Context*. New York: Bantam Books.

Flaste, Richard, ed. 1991. *The New York Times Book of Science Literacy*. New York: The New York Times Company.

Friedman, Herbert. 1990. *The Astronomer's Universe: Stars, Galaxies, and Cosmos*. New York: Ballantine Books.

Galton, Lawrence. 1985. *Med Tech: The Layperson's Guide to Today's Medical Miracles*. New York: Harper & Row, Publishers, Inc.

Golob, Richard, and Eric Brus, eds. 1990. *The Almanac of Science and Technology*. Boston: Harcourt Brace Jovanovich, Publishers.

Gribbon, John R. 1988. *The Omega Point: The Search for the Missing Mass and the Ultimate Fate of the Universe*. New York: Bantam Books.

Gribbon, John R. and Mary Gribbon. 1990. *Children of the Ice: Climate and Human Origins*. Cambridge, Mass.: Basil Blackwell Inc.

Hart, John Fraser. 1991. *The Land That Feeds Us*. New York: W. W. Norton & Company.

Hawking, Stephen W. 1988. *A Brief History of Time*. New York: Bantam Books.

Hazen, Robert M. 1988. *The Breakthrough*. New York: Summit Books.

Herbert, Nick. 1985. *Quantum Reality: Beyond the New Physics*. Garden City, N.Y.: Anchor Press/Doubleday.

Hofstadter, Douglas R. 1979. *Godel, Escher, Bach: An Eternal Golden Braid*. New York: Vintage Books.

———. 1986. *Metamagical Themas: Questing for the Essence of Mind and Pattern.* New York: Bantam Books.

Horner, John R., and James Gorman. 1988. *Digging Dinosaurs.* New York: Workman Publishing.

Jastrow, Robert. 1981. *The Enchanted Loom: Mind and the Universe.* New York: Simon & Schuster.

Jesperson, James, and Jane Fitz-Randolph. 1981. *From Quarks to Quasars.* New York: Macmillan Publishing Company.

Johnson, George. 1986. *Machinery of the Mind.* New York: Times Books.

Kurzweil, Raymond. 1990. *The Age of the Intelligent Machines.* Cambridge, Mass.: MIT Press.

Lafavore, Michael. 1987. *Radon: The Invisible Threat.* Emmaus, Pennsylvania: Rodale Press.

Lee, Thomas F. 1991. *The Human Genome Project: Cracking the Genetic Code of Life.* New York: Plenum Press.

Levenson, Thomas. 1989. *Ice Time: Climate, Science and Life on Earth.* New York: Harper & Row.

Lovelock, James. 1988. *The Ages of Gaia.* New York: W. W. Norton & Company.

Mallove, Eugene T. 1987. *The Quickening Universe: Cosmic Evolution and Human Destiny.* New York: St. Martin's Press.

———. 1991. *Fire and Ice: Searching for the Truth Behind the Cold Fusion Furor.* New York: John Wiley & Sons, Inc.

McKinnel, Robert Gilmore. 1985. *Cloning of Frogs, Mice and Other Animals.* Minneapolis: University of Minnesota Press.

Morriss, Richard. 1983. *Dismantling the Universe: The Nature of Scientific Discovery.* New York: Simon & Schuster.

———. 1984. *Time's Arrows: Scientific Attitudes Towards Time.* New York: Simon & Schuster.

Neiring, William A. *Wetlands: The Audubon Society Nature Guide.* New York: Alfred Knopf.

Nichols, Eve K. 1988. *Human Gene Therapy.* Cambridge, Mass.: Harvard University Press.

Parker, Barry R. 1990. *Colliding Galaxies.* New York: Plenum Press.

Preiss, Byron, and William R. Aschuler, eds. 1989. *The Microverse.* New York: Bantam Books.

Preston, Richard. 1987. *First Light.* New York: The Atlantic Monthly Press.

Raup, David M. 1986. *The Nemesis Affair: A Story of the Death of Dinosaurs and the Ways of Science.* New York: W. W. Norton & Company.

Restak, Richard M. 1989. *The Mind.* New York: Bantam Books.

Rheingold, Howard. 1991. *Virtual Reality.* New York: Summit Books.

Rhodes, Richard. 1986. *The Making of the Atom Bomb*. New York: Simon & Schuster.

Russell, Bertrand. 1958. *The ABC of Relativity*. New York: Signet Science Library Books.

Shapiro, Robert. 1991. *The Human Blueprint*. New York: St. Martin's Press.

Shipman, Harry L. 1976. *Black Holes, Quasars and the Universe*. Boston: Houghton Mifflin Company.

Sochurek, Howard. 1988. *Medicine's New Vision*. Easton, Pa.: Mack Publishing Company.

Sutton, Christine. 1984. *The Particle Connection*. Simon & Schuster.

Trefil, James. 1980. *From Atoms to Quarks*. New York: Charles Scribner's Sons.

———. 1983. *The Moment of Creation: Big Bang Physics from Before the First Millisecond to the Present Universe*. New York: Charles Scribner's Sons.

———. 1984. *A Scientist at the Seashore*. New York: Charles Scribner's Sons.

Tucker, Wallace, and Karen Tucker. 1988. *The Dark Matter*. New York: William Morrow and Company.

Ward, Peter Douglas. 1992. *On Methuselah's Trail: Living Fossils and the Great Extinctions*. New York: W. H. Freeman and Company.

Weart, Spencer R., and Melba Phillips, eds. 1985. *History of Physics*. New York: American Institute of Physics.

Wills, Christopher. 1991. *Exons, Introns, and Talking Genes: The Science Behind the Human Genome Project*. New York: Basic Books.

Wingerson, Lois. 1990. *Mapping Our Genes: The Genome Project and the Future of Medicine*. New York: Dutton.

Young, Louise B. 1986. *The Unfinished Universe*. New York: Simon & Schuster.

INDEX

ABOUT THE AUTHOR

Jo Ann Shroyer is an award-winning journalist who has earned a reputation for authoritative, accurate, and entertaining science writing. She was the science and medicine reporter for Minnesota Public Radio and her stories have aired on National Public Radio's "All Things Considered" and "Morning Edition." She has also written dozens of articles on astronomy, zoology, medicine, chemistry, and other scientific topics for numerous radio programs and journals. This is her first book.